WITHDRAWN

ECOLOGY AND ENERGETICS OF CONTRASTING SOCIAL SYSTEMS IN PHAINOPEPLA NITENS (AVES: PTILOGONATIDAE)

ECOLOGY AND ENERGETICS OF CONTRASTING SOCIAL SYSTEMS IN PHAINOPEPLA NITENS (AVES: PTILOGONATIDAE)

BY

GLENN E. WALSBERG

UNIVERSITY OF CALIFORNIA PRESS
BERKELEY • LOS ANGELES • LONDON
1977

UNIVERSITY OF CALIFORNIA PUBLICATIONS IN ZOOLOGY

Volume 108
Approved for publication June 11, 1976
Issued April 18, 1977

UNIVERSITY OF CALIFORNIA PRESS
BERKELEY AND LOS ANGELES
CALIFORNIA

◇

UNIVERSITY OF CALIFORNIA PRESS, LTD.
LONDON, ENGLAND

ISBN: 0-520-09562-6
LIBRARY OF CONGRESS CATALOG CARD NUMBER: 76-024475

COPYRIGHT © 1977 BY THE REGENTS OF THE UNIVERSITY OF CALIFORNIA
PRINTED IN THE UNITED STATES OF AMERICA

CONTENTS

Abstract	1
Introduction	2
Acknowledgments	2
General Biology of *Phainopepla Nitens*	3
Study Sites	4
Colorado Desert Washes	4
Riparian Woodland Areas	5
Methods	6
Resource Estimates	7
Time Budgets	8
Estimates of Daily Energy Expenditure	8
Colorado Desert System	13
Resources	13
Territoriality	18
Courtship	19
Breeding Behavior and Success	22
Riparian Woodland System	26
Resources	26
Territoriality	29
Social Behavior Associated with Food	32
Courtship	33
Breeding Behavior and Success	34
Time and Energy Budgets	34
Relation Between Oxygen Consumption and Ambient Temperature	37
Discussion	46
Ecology of Territorial Systems	46
Courtship Behavior	49
Influences of Resource Availability upon Breeding	51
Use and Evolution of Dual Breeding Ranges	54
Energetics	55
References	61

ECOLOGY AND ENERGETICS OF CONTRASTING SOCIAL SYSTEMS IN PHAINOPEPLA NITENS (AVES: PTILOGONATIDAE)

BY

GLENN E. WALSBERG

ABSTRACT

The Phainopepla, *Phainopepla nitens* (Ptilogonatidae), is a small (24 g) frugivorous and insectivorous bird that ranges from the southern portion of the Mexican Plateau into the southwestern United States. In California, Phainopeplas winter in the Colorado Desert, where they breed in March and April. They leave this area in late April and enter the coastal oak and riparian woodlands, where breeding occurs from late May through July. Data gathered in this study indicate that the same population breeds in both areas. This study analyzes this extremely unusual system of dual breeding ranges and deals with the following questions: (1) What are the important characteristics of the two habitats, particularly the distribution, abundance, and use of the major resources? (2) In what manner is the Phainopepla's social behavior adapted to the contrasting patterns of resource availability? (3) What are the consequences of the use of the two contrasting habitats and social systems for the bird's use of time and energy?

This study was conducted from 1972 to 1975. Intensive studies were made along washes in the Colorado Desert and in the riparian woodlands of the Santa Monica Mountains in coastal southern California.

General patterns of food use are similar in each area. Fruit is the adult Phainopepla's major food, while the young are also fed large amounts of insects. There are striking differences between the two habitats in the availability of these resources. Phainopeplas in the Colorado Desert feed primarily upon the berries of desert mistletoe (*Phoradendron californicum*), which parasitizes the large bushes and small trees that border desert washes. This mistletoe bears fruit from November through April and represents a stable food supply; individual mistletoe clumps require an average of seven weeks to decline to 50 percent of their peak number of berries measured. Mistletoe may be superabundant in December and January but declines greatly by the breeding period in March and April.

Insect abundance is very low in the Colorado Desert during the winter. It rises abruptly in the spring to about ten times winter levels, then declines.

This separation of the peak levels of insect and fruit abundance does not occur during the Phainopepla's summer breeding in the Santa Monica Mountains. Aerial insects are abundant throughout the summer and the Phainopepla's most important fruit in this area, that of *Rhamnus crocea,* peaks in abundance during the breeding period. This shrub occurs on chaparral-covered hillsides; it is isolated from the trees in the canyon bottoms which are used for nesting. The location of fruit is relatively unpredictable, since individual plants hold abundant fruit for only one to two weeks. The peak standing crop of *Rhamnus* berries is equal to only 7.5 percent of that of desert mistletoe berries.

Dramatic seasonal shifts were observed in territorial behavior, correlated with these contrasting resource patterns. Large ($\bar{x}$ = 0.40 ha) feeding and nesting territories are defended along desert washes. Territories are much smaller ($\bar{x}$ = 0.03 ha) and restricted to courtship and nesting in the Santa Monica Mountains, where coloniality is common. These territorial systems are analyzed as adaptations to the differing temporal and spatial patterns of food availability.

The Phainopepla's courtship behavior shows adaptations to a compressed breeding period. Such a breeding period is seen most clearly in the Colorado Desert, where abundant insects and sufficient fruit with which to feed young occur simultaneously for only six to eight weeks annually. Adaptations facilitating a rapid response to favorable breeding conditions include nest-building by the unmated male, which is apparently stimulated by the start of the desert's spring insect bloom, and the initiation of courtship as early as mid-January.

Differences observed between courtship in the two habitats include the omission in the woodlands of certain displays used conspicuously in the desert. These displays appear to facilitate courtship in the desert by reducing the male-female aggression that is residual from the establishment of winter territories. Such reduction of aggression is unnecessary in the riparian woodlands, since intersex territoriality never occurs there.

Significant effects of resource availability upon the timing and success of breeding were observed, including widespread breeding failures associated with low levels of fruit and insect abundance. Significant differences in clutch-size are correlated with differing levels of insect abundance.

Time budgets were calculated for Phainopeplas in both habitats. Associated with the contrasting social systems, the Phainopepla's use of time differed greatly between habitats, with woodland birds consistently spending much more time in foraging flights and social activity than desert birds.

Daily energy expenditure (DEE) was estimated using these time budgets. DEE varied from 12.74 to 22.27 kcal/day and changes in breeding status and habitat were associated with major differences in energy expenditure. The DEE of woodland birds was 19 to 28 percent higher than that of desert birds, primarily due to increased time spent in flight. The ecological significance of these differences in energy expenditure is discussed.

INTRODUCTION

Studies of avian social systems generally have dealt with either the single social system of a particular species or have correlated interspecific diversity in social systems with environmental variation. However, an individual bird species commonly inhabits a variety of ecological communities. Miller's (1951) analysis of California bird species revealed that 76 percent breed in two or more distinct ecological communities, and 45 percent breed in three or more. How is the social behavior of these birds influenced by, or adapted to, the differing environmental conditions these habitats present? Because of its virtually unique annual cycle, the Phainopepla (*Phainopepla nitens*) offers excellent opportunities for the study of this problem. In California, Phainopeplas winter mainly in the Colorado Desert, where they breed in the early spring. They then leave this area and appear in the coastal riparian and oak woodlands, where they breed during the summer. The intent of this study is to analyze this sequential residency and breeding in two dissimilar habitats, in order to answer the following questions:

1. What are the general characteristics of the Phainopepla's annual cycle? Especially, do individuals breed both in the Colorado Desert and later in the coastal woodlands or, as Crouch (1943) suggested, does a portion of the population breed in each habitat and remain as a nonbreeding element in the other?

2. What are the important characteristics of the two habitats; particularly the distribution, abundance, and use of major resources?

3. In what manner is the Phainopepla's social behavior influenced by, and adapted to, the contrasting patterns of resource availability?

4. What are the consequences of the use of the two contrasting habitats and social systems for the bird's use of time and energy?

Acknowledgments

I would like to thank T. R. Howell for his guidance throughout this study, as well as for his valuable comments on the manuscript. I also thank E. B. Edney and G. A. Bartholomew for critically reading the manuscript and N. E. Collias for valuable discussions of Phainopepla behavior. G. A. Bartholomew generously made laboratory facilities available. I am grateful to K. M. Bell-Walsberg for her field assistance throughout the course of this study. A number of my fellow graduate students aided significantly during this study, and I thank especially R. J. Epting, S. Hillman, D. F. Hoyt, and E. N. Mirsky for valuable discussions. R. J. Epting and D. F. Hoyt greatly aided in data analysis and P. Withers photographed the figures. L. Shields gave valuable field assistance and F. M. Fox aided in the collection and analysis of *Rhamnus* berries. Egg data from the Western Foundation of Vertebrate Zoology was kindly supplied by L. Kiff. I thank B. W. Anderson for freely sharing his data on Phainopepla populations in the Colorado River Valley and P. J. Gould for loaning me his unpublished manuscript on Phainopepla territoriality and courtship in Arizona. A portion of the fieldwork was done at the Phillip L. Boyd Deep Canyon Desert Research Center, and I thank J. Zabriskie and W. Jennings for their hospitality and assistance during my stay there. This study was supported by grants in 1974 and 1975 from the Frank M. Chapman Memorial Fund of the American Museum of Natural History and National Science Foundation grant GB 41277 in 1974.

GENERAL BIOLOGY OF *PHAINOPEPLA NITENS*

The Phainopepla is the most northern representative of the small family Ptilogonatidae (four species), which extends south into Central America (Greenway, 1960). *Phainopepla nitens* ranges from the Mexican Plateau near Puebla and Veracruz north into the southwestern United States; the race *P. nitens lepida* occupies almost all the species' range in the United States (American Ornithologists' Union, 1957). Phainopeplas in the United States winter primarily in the Sonoran Desert and shift in the spring to areas east, west, and north of that desert (fig. 1). Sufficient data are not available to describe seasonal movements in Mexico. In the southern portion of its

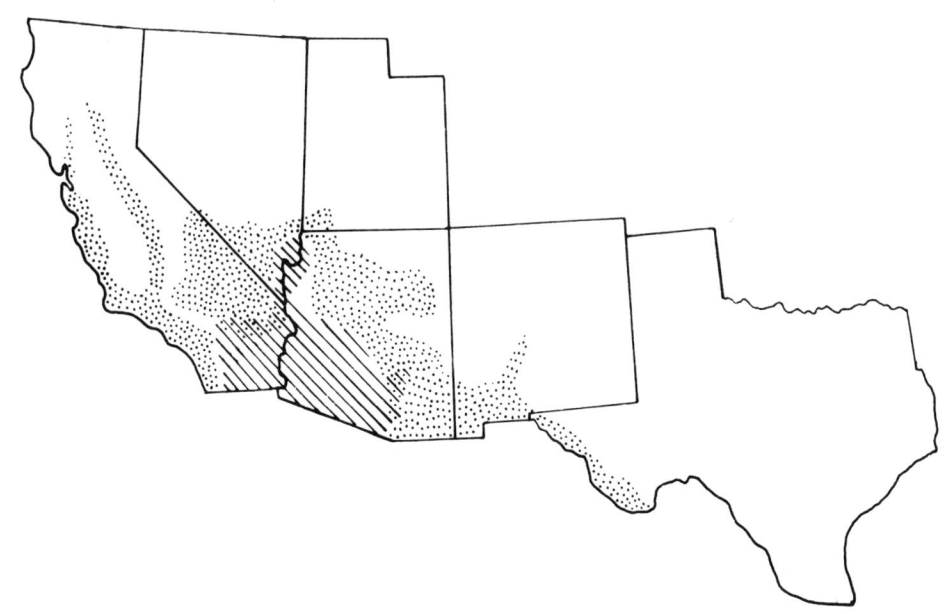

Fig. 1. Seasonal distribution of *Phainopepla nitens* in the southwestern United States. Hatched areas represent "winter" range, generally from October or November through April. Stippled areas represent "summer" range, generally from April until October or November. Based upon personal observations and Bailey, 1928; Crouch, 1943; Grinnell and Storer, 1924; Grinnell and Miller, 1944; Johnson et al., 1948; Miller and Stebbins, 1964; Phillips et al., 1964; and Wauer, 1964, 1969.

range in Mexico, the Phainopepla is generally associated with arid subtropical scrub. In the Sonoran Desert, farther north, it is usually restricted to communities that represent relict elements of this association (Axelrod, 1958). These communities contain a variety of arborescent plants and are usually restricted to more mesic areas in the desert, often along drainage systems that may flood after heavy rains or in regions with a locally high water table. Outside of the desert, Phainopeplas are found most commonly in oak and riparian woodlands and less commonly in a variety of other semiarid woodlands.

Phainopeplas are small birds, averaging 24.0 g (range, 22-28 g; n = 33), but are visually striking. The male is glossy black with white wing patches, whereas the female is dull gray with light gray wing patches. Both sexes have crests and red irides. Food consists both of insects, which are caught primarily by short, sallying flights from a perch, and a variety of small berries. In the Sonoran Desert, the major fruit fed upon is that of the desert mistletoe (*Phoradendron californicum*), and *P. nitens* is an important disseminator of this plant's seeds (Cowles, 1936, 1972; Crouch, 1939, 1943). Like a number of other mistletoe specialists, the Phainopepla possesses a highly modified stomach (Walsberg, 1975).

The major previous studies of Phainopepla social behavior were based primarily upon observations in the coastal mountains of San Diego County, California (Crouch, 1939) and in riparian associations near Tucson, Arizona (Rand and Rand, 1943). These authors described courtship and breeding behavior in these areas, including song, courtship chases, courtship feeding, nest-building by the unmated male, and the relatively equal division of incubation between the sexes. A number of the descriptions contradict one another, including those of the use of song in territoriality, courtship chases, nest-building by the unmated male, and the male's role in the care of the nestlings.

STUDY SITES

Colorado Desert Washes

The Colorado Desert is the most barren subdivision of the Sonoran Desert and includes all the Sonoran Desert in California. It is bounded on the east by the Colorado River, on the north by the Mojave Desert, on the west by the peninsular ranges, and in the south extends into Baja California (Jaeger, 1957). Much of this desert is near or below sea level and annual rainfall is typically less than 10 cm. Summers are hot, with air temperatures often reaching 40-48°C. Winters are mild. Temperatures occasionally reach 0°C, but even in the coldest months it usually warms during the day to 15-20°C. Thus, because of their residency in the Colorado Desert only from November to May, Phainopeplas experience relatively mild average temperatures (fig. 2). Wind storms are also a conspicuous feature of this desert, generally occurring one to three times per month. The storms last a few days, with winds from 30 to 80 km/hr.

All of my major desert study sites were washes—drainage channels through canyons, over alluvial fans, and through desert basins. These generally contain no surface water except once or twice a year, when they may flood after a heavy rain. The large bushes and small trees that typically border a wash create a belt of arborescent vegetation in

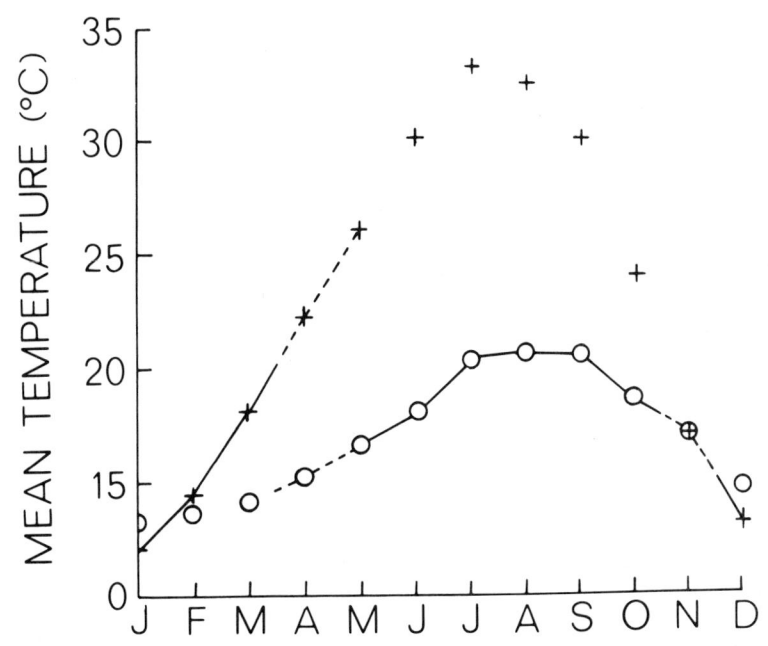

Fig. 2. Mean monthly temperatures in the Colorado Desert and in the Santa Monica Mountains. Crosses: temperatures in the Colorado Desert, as measured at Indio, California. Circles: temperatures at UCLA, at the base of the Santa Monica Mountains (U.S. Weather Bureau, 1964). Solid lines denote periods of residency in an area, dashed lines denote migratory periods.

an area that otherwise is usually dominated by creosote bush scrub. The dominants along a wash are usually some of the following species: catclaw acacia (*Acacia greggi*), mesquite (*Prosopis juliflora*), palo verde (*Cercidium floridum*), smoke tree (*Parosela spinosa*), ironwood (*Olneya tesota*), and desert willow (*Chilopsis linearis*). The surrounding creosote bush scrub is dominated by creosote bush (*Larrea divaricata*) and burro-weed (*Franseria dumosa*), with ocotillo (*Fouquieria splendens*), cholla (*Opuntia* spp.) and saltbush (*Atriplex* spp.) often abundant.

Observations were made at numerous points throughout the Colorado Desert, with intensive studies at three sites: (1) Milpitas Wash, 170 m elevation, 40 km southwest of Blythe, in Imperial County, California. (2) Shaver's Wash, 460 m elevation, 59 km east of Indio, Riverside County, California. (3) Coyote Wash, 120 to 330 m elevation, 7 km south of Palm Desert in the Philip L. Boyd Deep Canyon Desert Research Center, Riverside County, California.

Riparian Woodland Areas

All of my major woodland study sites were in the Santa Monica Mountains, which lie north and northwest of Los Angeles and form part of the outer coastal ranges of California. No part of the mountain range lies more than 15 km from the sea and the area experiences a "Mediterranean" climate: hot, dry summers and mild winters with rainfall averaging locally from 35 to 50 cm. Average temperatures in the area are within the range of fall, winter, and spring temperatures in the Colorado Desert, though daily variation is not as extreme (fig. 2). Riparian woodlands are located at the bottom of numerous canyons in these mountains; the surrounding hills are usually chaparral-covered. Observations were made in many of these canyons, with intensive studies at Decker Canyon and Rustic Canyon.

DECKER CANYON

This site is located in Ventura County, about one kilometer south of Westlake Village. Elevation of the canyon bottom is about 300 m. Located on the interior side of the mountains, this was the driest and hottest of my woodland sites. During the summer, temperatures commonly reach 40°C and the stream is reduced to a few pools. The canyon bottom is dominated by valley oak (*Quercus lobata*) and sycamore (*Platanus racemosa*), with a ground cover of grass and a few scattered shrubs—mostly poison oak (*Rhus diversiloba*) and laurel sumac (*Rhus laurina*). Hillsides are dominated by grass intermixed with patchy, low chaparral. The chaparral consists mainly of chamise (*Adenostoma fasciculatum*) with a scattering of other shrubs, primarily buckthorn (*Rhamnus crocea*), laurel sumac, and poison oak.

RUSTIC CANYON

This site is located in the portion of Rustic Canyon about three kilometers northwest of Sunset Boulevard, in Los Angeles County. Canyon bottom elevation is about 150-180 m. This site, on the coastal side of the mountains, is much wetter than Decker Canyon. Stream flow is continuous in June, but later in the summer sections of the stream dry up. The canyon bottom is dominated by coast live oak (*Quercus agrifolia*) and sycamore. The chaparral-covered hillsides are dominated by chamise, bigpod ceanothus (*Ceanothus megacarpus*), toyon (*Heteromeles arbutifolia*), and laurel sumac.

METHODS

This study was conducted between June 1972 and May 1975. Intensive studies in the Colorado Desert were made at Milpitas Wash from January through April 1974 and at Coyote Wash from March through April 1975. Periodic trips were also made to Colorado Desert sites between June 1972 and May 1973, November and December 1973, and November 1974 and February 1975. Intensive studies in the Santa Monica Mountains were made from April through August 1974. Additional observations were made between April and September 1973 and between June and September 1972.

For behavioral studies, Phainopeplas were captured with mist-nets, weighed to the nearest 0.1 g, and banded with plastic color bands and standard U.S. Fish and Wildlife Service bands. Egg data are from personal records and the Western Foundation of Vertebrate Zoology (WFVZ). Estimates of mean clutch size are based upon clutches known to be complete, and WFVZ records were used only if the collector had noted that incubation had been initiated. WFVZ records were used to supplement personal records of egg-laying dates if the collector had noted that the eggs were fresh.

Resource Estimates

FRUIT

Estimates of the standing crop of desert mistletoe berries (*Phoradendron californicum*) were made in three contiguous Phainopepla territories at Milpitas Wash. These territories included all mistletoe within a 100 × 250 m rectangle (2.5 ha; see fig. 4). At least 80 percent of the fruiting mistletoe clumps in each territory were accessible to me. Inaccessible clumps, usually at the outer edge of a tall tree, were assumed to contain the mean number of berries per accessible clump and the appropriate correction made. Desert mistletoe clumps, which consist of a mass of berries and leafless stems, are usually spheroidal, and fruiting is confined to an external shell 9-23 cm deep in large clumps. Major and minor diameters, as well as depth of the fruiting layer, were measured to the nearest centimeter. This allowed estimation of fruiting volume, assuming clumps approximate spheroids. Volumes of the few clumps not resembling spheroids were estimated using the equation for the geometric figure most closely approximated. A cylindrical, 500 ml jar was worked into the fruiting layer and that volume of the fruiting layer was removed. Berries were counted after being divided into three size classes based upon diameter (d): (1) $d \leqslant 3$ mm, (2) 3 mm $< d \leqslant 4$ mm, (3) $d > 4$ mm. (Mistletoe berries rarely exceed 6 mm in diameter.) Sampling was repeated every 14 days.

Estimates of *Rhamnus crocea* berries were made at Decker Canyon. All fruiting *Rhamnus* in a 27 ha area were mapped and *Rhamnus* in a 9 ha subsection were censused individually. Volume was estimated by measuring to the nearest centimeter the height, width, and depth of the fruiting portions of a shrub. Berry counts were made with the aid of a skeletal, cubic frame that had one end removed, allowing insertion into a shrub with minimal disturbance. Two sizes of frames were used. For low berry densities, the frame was 30 l in volume (31.1 cm on a side). For high berry densities, the frame was 10 l in volume (21.5 cm on a side). All berries within the frame were counted, but not removed. Fully red berries were counted as ripe and few berries appeared intermediate between ripe and nonripe. Counting was repeated weekly.

Caloric values for mistletoe berries are from Walsberg (1975). These values and values for *Rhamnus* were measured using a Phillipson Microbomb Calorimeter, with samples dried to constant weight at 75°C. Samples consisted of 82 *Rhamnus* berries and of 300 *Phoradendron* berries.

Census areas were mapped with a plane-table. Slope at Milpitas Wash averaged 1 percent and was ignored. Relief was added to maps of Santa Monica Mountain sites from enlargements of U.S. Geological Survey topographic maps.

INSECTS

Estimates of the diurnal population of aerial insects were made using a Malaise trap, which intercepts flying insects and funnels them into a cyanide-charged chamber. The trap was operated from sunrise to sunset, almost every day without high winds in the desert, weekly at Rustic Canyon and biweekly at Decker Canyon. Insects were removed daily, dried to constant weight and weighed to the nearest 0.1 mg.

Time Budgets

Time budgets were calculated for individuals in three breeding statuses: nonbreeding, incubating, and caring for nestlings. An individual was timed with stopwatches one-half of every hour, from the time it first became active until it went to roost for the night. I always remained at a distance such that the bird's behavior was not noticeably affected, usually at least 30 m. Since time-budgets of adults feeding nestlings probably change significantly during the nestling period, all such budgets were calculated five to six days posthatching, when the young reach approximately 50-60 percent adult weight (Crouch, 1939). Timed behavior was divided into seven categories, all except flight being mutually exclusive: (1) flight, (2) perching, (3) transit, including all locomotion when social behavior or flycatching was not involved, (4) eating fruit, (5) flycatching, including only the time actually spent in flight, (6) nest attentiveness, the time spent at a nest with eggs or young, (7) social activity, including courtship, aggression, territorial displays, and nest-construction. (Not included was territorial advertisement by calls and conspicuous perching, unless obviously stimulated by an intruder or potential intruder.) Knowing the time spent in flight, as well as the total time active per day, allows calculation of time spent in three activity level categories: (1) flight, (2) nonflight activity, equal to the bird's active day minus time in flight, (3) inactive, presumably asleep.

Estimates of Daily Energy Expenditure

Utter (1971) and Utter and LeFebvre (1973) compared estimates of daily energy expenditure (DEE) using time-budget methods and the more direct D_2O^{18} method and showed the following model to yield relatively accurate results:

$$\text{DEE} = (\dot{E}_f \times t_f) + (\dot{E}_{nf} \times t_{nf}) + (\dot{E}_s \times t_s) \tag{1}$$

where $\dot{E}_f$ is the rate of energy expenditure in flight, t_f is the time spent in flight per day, $\dot{E}_{nf}$ is the rate of energy expenditure during nonflight activity, t_{nf} is the time spent in nonflight activity per day, $\dot{E}_s$ is the rate of energy expenditure during the bird's inactive period (sleep), and t_s is the time spent inactive per day. I made this model more sensitive to the animal's temperature regime by incorporating two assumptions on the interaction of cold and exercise thermogenesis. First, exercise metabolism at the low work levels of nonflight activity is considered additive to maintenance metabolism at or below thermal-neutral temperatures, as indicated by the work of West and Hart (1966), Kontogiannis (1968), Pohl (1969), and Pohl and West (1973). Second, exercise metabolism at the high work levels of flight is assumed to substitute for cold-induced thermogenesis, as indicated by the work of Lasiewski (1963), Tucker (1968), and Berger and Hart (1972).

To reflect these assumptions, $\dot{E}_f$ is maintained constant (independent of temperature) and is assumed to incorporate maintenance metabolism. Temperature-dependent maintenance metabolism ($\dot{E}_m$) is estimated for the entire day minus time spent in flight. Birds are assumed to be at maintenance levels at night and at maintenance levels plus an activity increment during the nonflight portions of the active day. Thus, my revised form of equation (1) may be stated:

$$DEE = (\dot{E}_f \times t_f) + (\dot{E}_{nfa} \times t_{nf}) + (\dot{E}_m \times 24 - t_f) \qquad (2)$$

where $\dot{E}_{nfa}$ is the rate of energy expenditure due to nonflight activity that is additional to the rate of maintenance expenditure.

Two other general assumptions were made for estimates of DEE. All time-budgeted Phainopeplas were assumed to weight 24.0 g, the average of 33 wild-caught birds (range, 22-28 g). The caloric equivalent of consumed oxygen is assumed to be 4.8 kcal/liter.

ENERGY EXPENDITURE DURING FLIGHT ($\dot{E}_f$)

$\dot{E}_f$ was estimated using the least-squares regression line that Hart and Berger (1972) fitted to measured $\dot{E}_f$ in nine bird species. This equation predicts 2.989 kcal/hour as the cost of flight for a 24.0 g bird.

MAINTENANCE ENERGY EXPENDITURE ($\dot{E}_m$)

$\dot{E}_m$ was assumed to equal the minimal metabolic rate of a fasting bird resting at a specified ambient temperature. For measurement of $\dot{E}_m$, Phainopeplas were captured and maintained as described by Walsberg (1975). Oxygen consumption was measured in an open-flow system using a Beckman G-2 Oxygen Analyzer. Metabolic chambers were made from gallon jars with ports fitted for air flow into and out of the chamber and for a thermocouple used to monitor chamber temperatures. Flow rate was 600 cc/min. at chamber temperatures of 5 to 35°C, 800 cc/min. at 40°C, and 1500 cc/min. at 43 to 45°C. These flow rates maintained chamber humidity below 8 g H_2O/m^3. A controlled temperature box maintained chamber temperature within 0.5°C. All runs were made in the dark on postabsorptive birds, one to four hours after the initiation of the light cycle of their day. Minimal oxygen consumption was assumed to equal the average of the lowest three readings of a bird that had been in the chamber at least 30 minutes and was corrected to standard temperature and pressure. $\dot{E}_m$ at a particular temperature was estimated using a least-squares regression fitted to metabolic data below the thermal-neutral zone, assuming 4.8 kcal/1 O_2 consumed (see section on the relation of oxygen consumption to ambient temperature).

An average daily temperature can be used to predict an average maintenance metabolic rate after correction for two sources of error. First, an average daily temperature includes temperatures of periods when the bird is in flight and maintenance metabolism is not temperature-dependent. Second, during periods of the day when ambient temperature exceeds the lower critical temperature (T_{lc}), maintenance metabolism is not predicted by the equation of the line below thermal neutrality. This involved only the bird's active period, since nocturnal temperatures never exceeded the T_{lc}.

The following measures and relations were used to correct for these errors. Temperatures were measured two ways, but always two meters above ground and to the nearest degree centigrade. Twenty-four hour maximum and minimum temperatures were measured in the shade with a Taylor maximum-minimum thermometer. The average of daily maximum and minimum temperatures was used as a close estimate of the true average daily temperature (Kincer, 1941). Shade air temperatures were also recorded hourly at the beginning of each time-budget period, allowing a separate estimation of the average active-period temperature (Kincer, 1941).

For the first correction, all hourly temperatures during the active period that were above the T_{lc} were set equal to it. I define the T_{lc} as that point where the regression line fitted to data below thermal neutrality intersects the level of mean minimal metabolism in the thermal neutral zone (= 29°C, see section on the relation of oxygen consumption to ambient temperature).

The second correction was to base the average daily temperature upon nonflight portions of the bird's day. This was done by weighting the temperature taken at the beginning of each time-budget period (T) by the proportion of time not spent in flight during that period (P). Thus, the effective temperature for calculation of maintenance metabolism during the bird's active period ($T_{m,ap}$) is the average of each of these weighted temperatures:

$$T_{m,ap} = \frac{\sum_{i=1}^{n} (T_i \times P_i)}{t_{nf}} \quad (3)$$

where n is the number of time-budget periods and t_{nf} is the time, in hours per day, spent in nonflight activity.

The average daily temperature (T_{ad}) must be a function of the temperatures of the bird's active period (T_{ap}) and its inactive period (T_{ip}):

$$T_{ad} = \frac{(T_{ap} \times t_{ap}) + (T_{ip} \times t_{ip})}{24} \quad (4)$$

where T_{ad} is estimated as (maximum + minimum)/2, T_{ap} is estimated as the average of temperatures taken hourly at the beginning of each time-budget period, t_{ap} is the time spent active per day (in hours), and t_{ip} is the time spent inactive per day (in hours). By rearranging equation 4, the function ($T_{ip} \times t_{ip}$) can be calculated, since T_{ad}, T_{ap} and t_{ap} are known:

$$T_{ip} \times t_{ip} = (T_{ad} \times 24) - (T_{ap} \times t_{ap}) \quad (5)$$

Combining this value for the inactive period with the corrected value from equation 3 for the active period ($T_{m,ap}$) allows calculation of a temperature (T_m) that predicts the average $\dot{E}_m$ for nonflight periods:

$$T_m = \frac{(T_{ip} \times t_{ip}) + (T_{m,ap} \times t_{nf})}{24 - t_f} \quad (6)$$

where t_f is the time spent in flight per day (in hours).

Because the Phainopepla is a black, desert-dwelling bird, it seemed particularly appropriate to estimate the impact of solar radiation upon its energy expenditure. Since a bird can rapidly and significantly alter its exposure to radiation by postural changes and movement into variable amounts of shade, an exact estimate was not attempted. Instead, a range of effects was estimated which should bracket the actual value. The effect of radiation upon the bird was estimated using a black-bulb thermometer made by inserting a thermometer into the center of a 38 mm table-tennis ball painted with total solar absorption paint. Readings were taken two meters above open ground after the thermometer had been exposed at least 30 minutes, at the same time shade air temperature was recorded (hourly at the initiation of each time-budget period).

For each 30 minute time-budget period, the percent of time the bird spent in full sun was estimated. If time in full sun exceeded a certain level, then the black-bulb temperature (T_{bb}) was assumed to represent the effective temperature gradient the bird experienced during that period. T_{bb} was then used in preference to that period's shade air temperature (T_a) in the calculation of T_m. Two models were used: (1) $T_{m,50}$: T_{bb} was used in preference to T_a for each period in which the bird was in full sun at least 50 percent of the time, (2) $T_{m,25}$: T_{bb} was used in preference to T_a for each period in which the bird was in full sun at least 25 percent of the time.

Simultaneous measurements at the Deep Canyon Desert Research Center of solar radiation (measured with a Weather Measure Corp. mechanical pyranograph), T_{bb}, and T_a allowed a test of the relation between the increases of T_{bb} over T_a and useful energetic input to the bird. In figure 3, these values are compared with the measured effects of simulated solar radiation upon two small, dark-colored birds: Cowbirds, *Molothrus ater*, (about 35 g) and blackened Zebra Finches, *Poephilis castanotis*, (10 - 13 g). Cowbirds exposed to radiation of 0.9 cal cm^{-2} min^{-1} showed a decrease in standard metabolism equivalent to a 10°C increase in air temperature (Lustick, 1969).

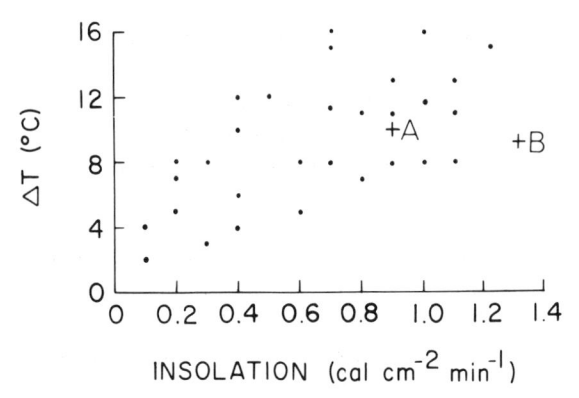

Fig. 3. The relation between solar radiation and the increase of black-bulb temperature over shade air temperature (ΔT). (A) represents the measured effect of simulated solar radiation upon Cowbirds, *Molothrus ater* (Lustick, 1969) and (B) represents the effect upon blackened Zebra Finches, *Poephilis castanotis* (Hamilton and Heppner, 1967). See text for methods and discussion.

Using conductance values derived from Calder (1964), I calculated that the decrease in standard metabolism of blackened Zebra Finches exposed to radiation of 1.31 cal cm^{-2} min^{-1} (Hamilton and Heppner, 1967) is equivalent to a 9.3°C increase in air temperature (range of averages for individual birds: 8.3 - 10.5°C). Comparison of these values with those for the increase of T_{bb} over T_a at various radiation levels indicates a generally close correspondence, with T_{bb} perhaps overestimating the effect of solar radiation (fig. 3).

ENERGY EXPENDITURE DUE TO NONFLIGHT ACTIVITY ($\dot{E}_{nfa}$)

Previously (Walsberg, 1975), I measured the energy expenditure over a 24-hour period of four Phainopeplas living individually in 93 × 60 × 45 cm cages on an 11-hour light, 13-hour dark photoperiod. These cages are large enough to permit normal activities such as feeding or hopping, but too small for significant amounts of flight. The intensity of nonflight activity in these cages and in the field appeared to be similar. Activity consisted mainly of quiet perching plus small amounts of hopping, singing, and feeding. The energetic cost of this nonflight activity was calculated by subtracting maintenance costs. Average energy expenditure for caged birds was 10.90 kcal/day or 0.0170 kcal g^{-1} hr^{-1} ($\bar{x}$ body weight = 26.7 g). The birds were kept at 25°C and maintenance metabolism was thus equal to 0.0141 kcal g^{-1} hr^{-1} (see section on the relation of oxygen consumption to ambient temperature). Therefore, the average cost of nonflight activity was 0.0029 kcal g^{-1} hr^{-1} (= 0.0170-0.0141). Concentrated into the 11-hour light portion of the caged birds' daily cycle, $\dot{E}_{nfa}$ for a bird weighing 24 g (the assumed weight of time-budgeted birds) is 0.152 kcal/hour [= 24 g × (24/11) × 0.0029 kcal g^{-1} hr^{-1}]. This is equal to 54 percent of *P. nitens'* basal metabolic rate (BMR), so that a bird in nonflight activity in thermal neutrality has an energy expenditure equal to 1.54 times BMR. This value is close to those derived independently by a number of authors, which range from 1.5 to 2.0 times BMR (Gessaman, 1973).

ENERGY EXPENDITURE REQUIRED FOR INCUBATION

The energetic cost of incubation is controversial. Kendeigh (1973) believes that a parent must significantly increase its energy expenditure over resting levels to supply heat required for incubation. King (1973) disagrees and believes that heat production as a by-product of the parent's resting metabolism can supply all, or a large portion, of the heat necessary for incubation. Though empirical data are lacking, it seems unlikely that incubation entails significant elevations in energy expenditure for the Phainopepla, due to its small clutch (2-3 eggs) and the moderate ambient temperatures during its incubation period. Thus, I have accepted King's (1973) view, and assumed that levels of energy expenditure estimated for maintenance metabolism and nonflight activity include costs related to incubation.

TEMPERATURE AND DAYLENGTH COMPENSATED MODEL

To estimate differences in energy expenditure due to social systems and resource patterns, differences in the two major physical influences of environmental temperature and daylength were approximately compensated for. All DEE estimates were recalculated using values for these factors typical of Phainopeplas breeding in the Santa Monica Mountains.

An average temperature of 20°C was assumed. Though this crude compensation affects only maintenance metabolism, a temperature change probably would also produce a change in the bird's time budget. Most importantly, birds during the winter and spring in the Colorado Desert exposed to the relatively warm temperature of 20°C would probably decrease time foraging and, hence, in flight. Thus, this compensation may overestimate DEE slightly for Colorado Desert birds.

The active day was assumed to equal 14.5 hours, and the bird was assumed to divide this period into the actually measured proportion of flight and nonflight activity. Here, a major error is that flights to food should be a function of DEE, and only indirectly a function of daylength. This model assumes foraging flights to be a direct function of daylength and probably overestimates flight time, and therefore DEE, in the desert.

COLORADO DESERT SYSTEM

Resources

GENERAL PATTERNS OF FOOD USE

Berries of the desert mistletoe (*Phoradendron californicum*) are the adult Phainopepla's major food in the Colorado Desert. Resident Phainopeplas are always associated with mistletoe, and local population densities usually reflect the local abundance of mistletoe berries. All sizes of berries are not fed upon equally; almost all berries eaten are in the medium size class (3-4 mm diameter). This preference is not solely a function of ripeness. While smaller berries (diameter < 3mm) are often not ripe, a few mistletoe clumps bear many large ripe berries (diameter > 6mm) that are ignored by Phainopeplas. A mean of 6.1 percent of the active day was spent eating mistletoe, with a tendency for birds to spend more time eating fruit during the winter than during the spring. Hence, the time birds spent eating fruit averaged 11.5 percent of the active day (1.1 hours) in January but only 4.7 percent (0.55 hours) in March (table 1, birds feeding nestlings not included).

Extensive flycatching, averaging 11.6 percent of the active day (1.43 hours), was seen only in birds feeding young (table 1). This is about 20 times the value for Phainopeplas without young, which averaged 0.62 percent of their active day (0.07 hours) flycatching.

DISTRIBUTION AND ABUNDANCE OF FRUIT

Desert mistletoe is a common parasite upon a number of arborescent legumes in the Colorado Desert, including palo verde, ironwood, mesquite, and catclaw acacia. These trees, which are also preferred nesting sites for Phainopeplas, commonly form a belt along desert washes (fig. 4), resulting in a rather uniform distribution of mistletoe clumps along a wash (fig. 9).

TABLE 1
Time Budgets of Phainopeplas: Behavioral Categories

No.	Site[1]	Status[2]	Sex[3]	Date	Flying	Perching	In transit	Eating fruit	Flycatching	At nest	In social activity
1	CD	NB	M	Jan. 23	11.5	83.0	1.5	11.0	0	—	10.0
2	CD	NB	M	Jan. 24	2.1	85.6	2.3	12.1	0.03	—	0
3	CD	NB	M	Feb. 13	1.5	90.1	1.1	8.4	0.02	—	0.3
4	CD	NB	M	Mar. 14	9.4	84.9	2.2	5.8	4.3	—	2.8
5	CD	NB	F	Jan. 29	2.1	85.4	1.6	12.0	0.04	—	0.9
6	CD	NB	F	Feb. 21	5.7	85.0	2.5	9.5	0.1	—	2.9
7	CD	NB	F	Feb. 24	5.3	90.7	2.1	3.7	0.03	—	3.5
8	CD	Inc.	M	Mar. 17	3.2	39.8	3.0	4.3	0.2	51.0	1.7
9	CD	Inc.	M	Mar. 21	2.9	53.4	2.8	3.9	0.8	38.0	1.2
10	CD	Inc.	F	Mar. 17	3.8	38.5	3.3	5.6	0.1	51.1	1.5
11	CD	Inc.	F	Mar. 21	3.2	36.8	2.4	3.7	1.8	54.2	1.1
12	CD	Nestling	M	Apr. 14	14.1	72.6	2.3	4.3	9.7	8.4	2.7
13	CD	Nestling	M	Apr. 15	16.4	71.1	2.7	4.0	11.0	8.2	3.1
14	CD	Nestling	M	Apr. 16	12.2	73.9	3.1	4.5	8.5	8.6	1.4
15	CD	Nestling	F	Apr. 14	16.3	68.9	2.1	4.6	13.6	8.8	2.0
16	CD	Nestling	F	Apr. 15	14.2	70.9	2.4	4.6	12.2	9.7	1.1
17	CD	Nestling	F	Apr. 16	17.2	66.4	2.8	5.0	14.6	9.1	2.1
18	SMM	NB	M	May 23	17.0	74.0	12.0	4.5	0.02	—	9.4
19	SMM	NB	M	May 29	16.4	70.8	13.0	5.0	0.08	—	11.1
20	SMM	NB	M	June 4	17.7	74.1	13.3	4.0	0.4	—	8.2
21	SMM	Inc.	M	June 12	12.9	40.7	11.7	4.2	0.1	41.0	2.3
22	SMM	Inc.	M	June 16	12.9	35.2	12.1	3.2	0.08	46.1	3.4
23	SMM	Inc.	M	June 17	13.6	37.9	13.1	3.6	1.0	42.7	1.8
24	SMM	Inc.	F	June 10	14.0	34.1	12.8	4.3	1.0	46.9	0.9
25	SMM	Inc.	F	June 15	13.7	31.8	14.0	4.7	0.8	44.8	3.9
26	SMM	Inc.	F	June 6	12.5	30.2	12.3	4.2	0.1	47.2	6.0

TABLE 1 — Continued

No.	Site[1]	Status[2]	Sex[3]	Date	Flying	Perching	In transit	Eating fruit	Flycatching	At nest	In social activity
27	SMM	Nestling	M	June 21	21.1	63.0	13.0	3.9	8.7	7.9	3.5
28	SMM	Nestling	M	June 28	23.0	65.1	11.9	3.6	9.9	7.4	2.1
29	SMM	Nestling	M	July 3	28.9	58.0	14.1	5.0	11.5	7.8	3.6
30	SMM	Nestling	F	June 22	21.5	54.5	15.7	4.1	7.6	8.0	10.1
31	SMM	Nestling	F	June 29	27.1	54.0	16.1	4.7	10.1	8.9	6.2
32	SMM	Nestling	F	July 4	24.2	48.7	16.7	4.8	12.2	8.2	9.4

[1] CD = Colorado Desert; SMM = Santa Monica Mountains.
[2] NB = nonbreeding; Inc. = incubating; Nestling = feeding and brooding young in nest.
[3] M = male; F = female.

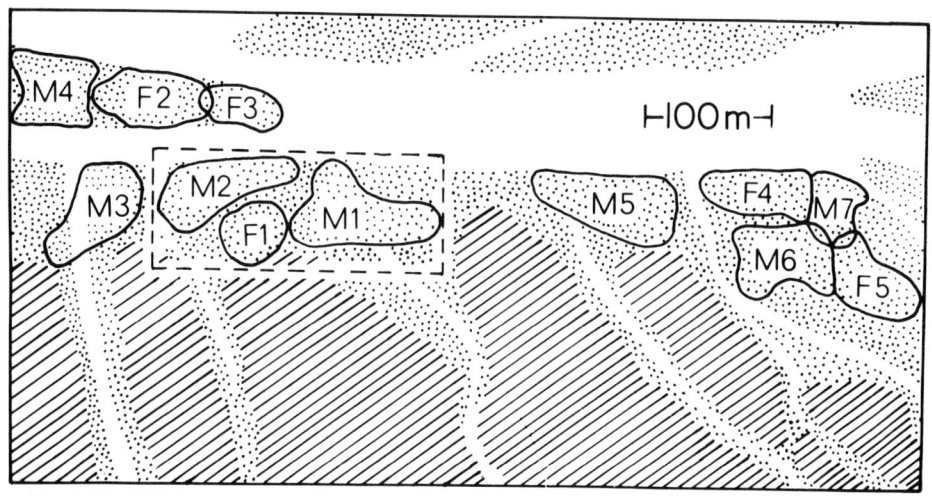

Fig. 4. The Milpitas Wash site. Blank areas are primarily nonvegetated sand, hatched areas are dominated by creosote bush scrub, and stippled areas are dominated by trees and large shrubs. Dashed line circumscribes the 2.5 ha mistletoe census area. Twelve mapped Phainopepla territories are shown as they existed prior to pairing (mid-January). Letter and number identify resident Phainopepla (M = male, F = female). Though only mapped on one side of the wash, territories were spread along both sides.

This mistletoe bears fruit in the Colorado Desert mainly from November through April, with the highest abundance reached in December or January. At the time of the first mistletoe census on January 31, 1974, all mistletoe clumps in the 2.5 ha census area (fig. 4) that would bear ripe fruit that season were already at their maximum fruit density measured, ranging from 60 to 540 ripe 3-4 mm berries per liter of the fruiting portion of the clump ($\bar{x}$ = 204). Clumps held fruit for long periods, taking an average of seven weeks to decline to one-half of their peak number of berries measured. This decline in berry abundance appeared to a large extent to be a result of fruit spoilage. Since fruit production was so high, Phainopeplas consumed only a portion of the crop before many ripe berries became overripe and dehydrated. Entire clumps of such berries were commonly seen.

Standing crop evidently peaked prior to the January 31 census and declined continually after that (fig. 5). In the 2.5 ha area, there were 71 clumps that held ripe fruit, of which 62 were actually censused. These clumps were up to 2.0 m wide and the volume of the fruiting portion averaged 49 liters/clump. On January 31, the 71 held about 716,000 ripe 3-4 mm berries, or an average of 286,000 per hectare (fig. 5). Assuming 83.9 cal/berry (Walsberg, 1975), this is equivalent to 24,000 kcal/ha. By March 14, the average date of egg-laying in the Colorado Desert (see section on breeding behavior), standing crop had declined 37 percent to 178,000 berries/ha (15,000 kcal). On the final April 11 census, it had further declined to about 55,000 berries/ha (4,600 kcal).

The three censused territories varied greatly in mistletoe abundance (fig. 6). Of the total standing crop in the three territories, the richest territory held 64 percent on January 31 and 96 percent on April 11.

Abundance of mistletoe fruit also varies greatly from year to year. Abundance in the seasons of 1971-1972 and 1972-1973 appeared to be as high as the 1973-1974 season,

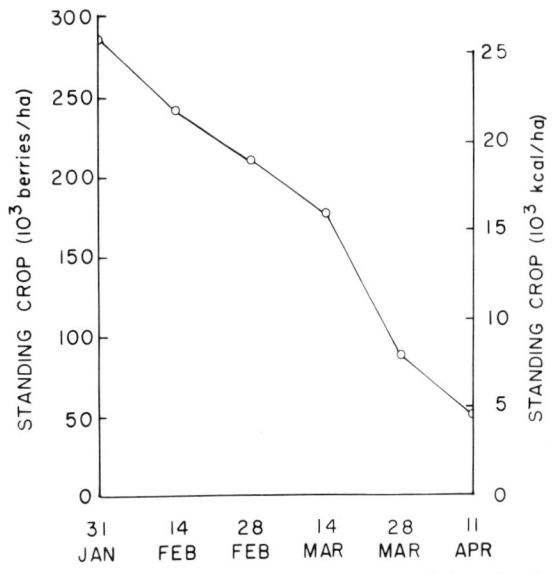

Fig. 5. Standing crop of ripe 3-4 mm desert mistletoe berries during an 11 week period at Milpitas Wash in 1974. Energy content calculated assuming 83.9 cal/berry (Walsberg, 1975).

when the above censuses were made. Compared to these years, mistletoe fruit was sparse throughout the Colorado Desert in the 1974-1975 season. Though not quantified, it appeared to be about one-third of the previous year's levels. Some areas, such as Shaver's Wash, totally lacked ripe fruit by the end of February.

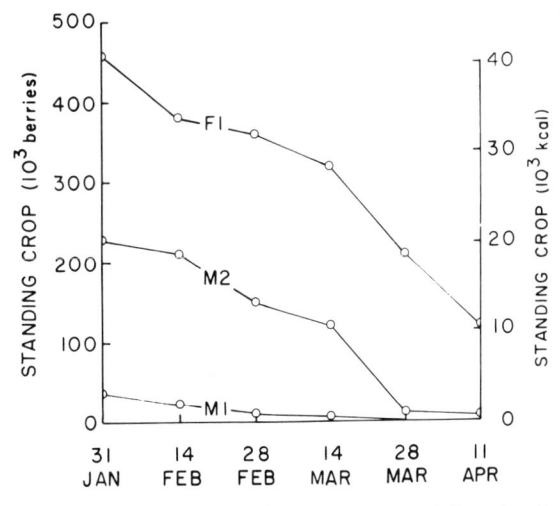

Fig. 6. Standing crop of ripe 3-4 mm desert mistletoe berries in three territories during an 11 week period at Milpitas Wash in 1974. Energy content calculated assuming 83.9 cal/berry (Walsberg, 1975).

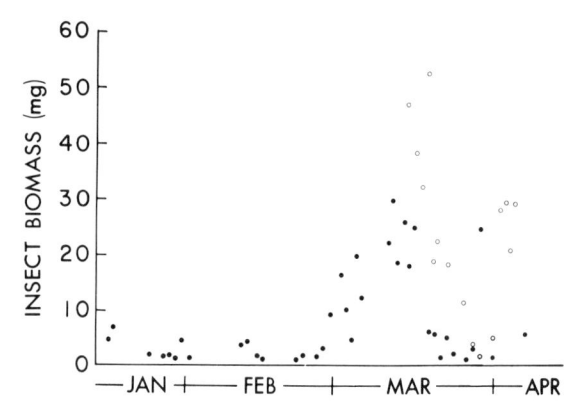

Fig. 7. Aerial insect abundance at Milpitas Wash in 1974 (closed circles) and Coyote Wash in 1975 (open circles). Data are dry weights of insects caught per day in Malaise trap.

INSECTS

The population of diurnal aerial-insects was censused at Milpitas Wash in 1974 and Coyote Wash in 1975 (fig. 7). At Milpitas Wash, between January 17 and February 27, an average of 2.4 mg/day (dry weight) of insects were caught. The spring insect bloom, first detected on February 28, peaked between March 11 and March 16, when an average of 23.2 mg/day (dry weight) of insects were caught (fig. 7). After March 16, insect abundance dropped to prebloom levels. This spring insect bloom appeared to coincide with the peak of the annual plant bloom.

These results from Milpitas Wash were roughly paralleled at Coyote Wash in 1975 (fig. 7). An average of 19.2 mg/day (dry weight) of insects were caught between March 15 and April 5, or about 2.4 times higher than the same period at Milpitas Wash the previous year. Again, the drop in insect abundance coincided with the end of the spring annual plant bloom.

Territoriality

CHARACTERISTICS OF TERRITORIES

Phainopeplas are territorial during their entire residency in the Colorado Desert, with territories usually spread along a wash in the belt of small trees and large shrubs (fig. 4). The territories of 12 color-banded birds were mapped at Milpitas Wash and averaged 0.377 ha (range, 0.244-0.604 ha). At Milpitas Wash, territories were stable by December 13, 1973. Males and females defend separate territories until courtship intensifies in late January and February, when territorial shifts become common. A male and a female that pair and had had adjoining territories, simply combine them. If the territories were not adjacent, then the pair usually occupies the male's territory.

One bird at Milpitas Wash (male 4 in fig. 4) held two territories after February 1. The second was gained by vigorously courting a female (female 1 in fig. 4), who avoided him and subsequently abandoned her territory. After successfully defending this area against neighbors, the male commuted between this territory and his original territory, usually spending a few hours to a day at a time in each.

The stability of territories at Milpitas Wash was associated with an abundant fruit supply. Major population shifts with the abandonment of territories accompanied the widespread failure of the mistletoe crop in 1975. Between December 1974 and January 1975, about one-half of the Phainopeplas left the mesquite association along the Colorado River near Blythe, California (B. Anderson, pers. comm.). This was apparently associated with a mistletoe crop failure. Later, between January 23 and February 7, the population at Coyote Wash approximately tripled, apparently due to an influx of Phainopeplas from elsewhere in the desert. Newly arriving birds immediately established territories and remained until April.

In addition to the intraspecific defense of a geographic area, Phainopeplas also defend fruiting mistletoe against almost all species that feed on it, with varying success. Gambel's quail (*Lophortyx gambelii*), sage thrashers (*Oreoscoptes montanus*), robins (*Turdus migratorius*), house finches (*Carpodacus mexicanus*) and white-crowned sparrows (*Zonotrichia leucophrys*) usually are quickly chased away. Bluebirds (*Sialis mexicanus* and *S. currucoides*) are so tenacious that Phainopeplas may chase individuals in circles around a mistletoe-bearing tree for up to 30 minutes, the bluebird returning to feed as soon as the Phainopepla perches. Bluebird flocks are often impossible to dislodge. Mockingbirds (*Mimus polyglottus*) also defend mistletoe fruit and are dominant over Phainopeplas. However, mockingbirds are successful only in harassing Phainopeplas and do not exclude them for more than a few minutes at a time from a mistletoe clump.

TERRITORIAL ADVERTISEMENT AND DEFENSE

Advertisement consists primarily of conspicuous perching, amplified by short call-notes. From the time of first activity until 0900, 80 percent of a nonbreeding bird's time is spent perched and 96 percent of the perching is done on top of a tree or snag. After courtship is initiated in January, courtship flights are also used in territorial display and defense (see section on courtship behavior).

Active defense usually consists of chasing an intruder from the territory, but boundary disputes sometimes take the form of "silent confrontations." These occur soon after two males oppose each other in a courtship flight or when a bird establishes itself adjacent to another territory. The two birds meet in a shrub or tree on their joint territorial border and sit silently, 50-70 cm apart. A variety of positions are held for up to 15 minutes. The birds may conspicuously face directly toward or away from each other, or parallel or perpendicular to each other. Except for occasionally shifting rapidly and completely between two positions, the birds perch rigidly. These confrontations may be maintained up to 30 minutes after all other Phainopeplas have gone to roost for the night. They end when one bird suddenly hops toward the other, who retreats, or when one bird leaves suddenly and is usually chased a short distance by the other.

COURTSHIP

The Phainopepla's residency in the Colorado Desert can be divided into three periods relative to courtship: (1) the precourtship period, (2) prenesting courtship, and (3) the nest-construction period. The precourtship period extends from the November entry into the Colorado Desert through January. Individuals are usually alone on their territories, and song in December and January is heard only during the last hours of daylight. Interactions appearing to be incipient courtship are seen as early as mid-December. During these interactions, a bird flies into a shrub or tree where there is a bird of the opposite sex. The birds hop through the vegetation together, then one flies off. These interactions are not frequently seen, but appear identical to a form of courtship seen later in the season.

Prenesting courtship at Milpitas Wash in 1974 was initiated about January 31 (when daylength was 10.6 hours) and lasted through February. A number of phenomena distinguishing this period from the previous one first occurred on January 31 and February 1, including the first courtship flights, courtship assemblages, transient pair-bonds, and excursions of birds off their territories. More gradual change was heard in song, which was restricted to late afternoon during the first week of February, but occurred throughout the day by the end of the third week. The following are conspicuous features of prenesting courtship.

COURTSHIP FLIGHTS

Courtship flights are seen from February until the spring migration and almost always occur near sunrise or sunset. Of 13 flights observed at Milpitas Wash, 11 occurred an average of 37 minutes prior to sunset and two occurred less than 45 minutes after sunrise. Typically, this display consists of a male flying from his perch to a height of 25 to 100 m over his territory, then flying in a circling or zigzag pattern within a 15 to 25 m wide area. Lasting from 15 seconds to 7 minutes, this display is occasionally performed by a male alone, but usually he is joined by one to nine Phainopeplas of both sexes. Generally, the birds do not chase or follow one another, but fly in separate paths in the same area. If only two males are involved, they usually fly in parallel patterns aligned with their territorial boundary, each bird on his own side. Courtship flights are conspicuously silent, except for an occasional harsh and trilled call-note heard in displays with two males.

These displays can apparently be stimulated by a female flying near a male. I instigated a number of them by flushing a female off her territorial perch at dusk. Males seeing her would rise into a display flight.

Courtship flights end in one of three ways. The birds may separate into their own territories with no further interaction. Alternatively, two opposing males may land next to each other on a territorial border, resulting in a "silent confrontation" (see section on territoriality), or a male and female may drop down into one of their territories, resulting in a transient pairing (see below).

EXCURSIONS OF INDIVIDUALS OFF THEIR TERRITORIES

During prenesting courtship, birds leave their territories for extended periods. This occurs infrequently in males and is usually associated with a courtship assemblage or courtship flight in a neighboring territory. In contrast, females are frequently off their territories for one-half to four hours at a time. They are apparently attracted across great distances by the sight of a male displaying or a courtship assemblage. A number of times I saw a female leave her territory and fly toward a male displaying more than 300 m away. Often these females settled in the territory of the displaying male or that of an interjacent male. The female may consort with a male in a manner identical to that previously described as incipient courtship, or a courtship assemblage may be evoked (see below), or a transient pairing may occur (see below).

COURTSHIP ASSEMBLAGES

Courtship assemblages are seen from early February through April and consist of 4 to 14 birds, usually with a male-female ratio of 2:1. They resemble a compact flock in a tree, excitedly feeding on mistletoe, displacing members of their own sex, and with males chasing females. These assemblages are frequently initiated by a female entering a male's territory, sometimes in response to his display to another female. This apparently attracts other Phainopeplas of both sexes. An assemblage breaks up after 5 to 20 minutes and a female new to the area often remains perched in a local male's territory, forming a transient or, occasionally, a stable pair (see below).

COURTSHIP CHASES

Courtship chases, in which a male chases a female from tree to tree, occur from early February through April. In pairs with adjoining territories, the female commonly enters the male's territory and is chased as she retreats back to her territory. She then returns and the performance is repeated. Call-notes are frequent and chasing may continue for 10 to 15 minutes. At Milpitas Wash in early February, the female often ended this by remaining in her territory. After mid-February, courtship feeding was often seen after these chases.

COURTSHIP FEEDING

Courtship feeding is seen from February until migration and typically is initiated by the female posturing in front of a male in a manner similar to a begging juvenile: crouching, probing at the male's bill, and often with fluttering wings. The male then regurgitates one to five berries and feeds her. Rarely do males feed insects to females.

TRANSIENT PAIRS

Transient pairings were first seen at Milpitas Wash on January 31. They occur when a female appears in a male's territory, either unprovoked, or after a courtship flight or courtship assemblage, or when a male has expanded his territory into that of a neighboring female. Instead of activity such as vigorous chasing, the male and female perch quietly 40 to 90 cm apart and follow one another loosely around the territory. After five minutes to four hours, the female abruptly leaves the male and often flies more than 500 m away, sometimes directly to another male's territory. Banded individuals were observed associating repeatedly in this manner with up to four potential mates.

STABLE PAIRS

Pairs that proved to be stable, in the sense that they persisted a number of days, were first seen at Milpitas Wash on February 16. Some pairs were maintained until migration in April, but many disintegrated after 2 to 14 days. Thus, an individual at Milpitas Wash commonly had three or four mates sequentially between February and April.

The nest-construction period, the third stage in courtship, started at the end of February at Milpitas Wash in 1974. The first nest-display flight was seen on February 26, when a male displayed to a nest from the previous year. All males at Milpitas Wash initiated nests between February 28 and March 6, whether they had a mate or not. Except for nest construction and display, behavior during this period is similar to prenesting courtship.

Nest display is seen in the Colorado Desert throughout March and April, and is performed commonly by the male, rarely by the female. The bird flies slowly to the nest-site, often with nest-material in its bill, with its tail depressed about 30° and the white wing-patches made especially conspicuous. Males use this display to lead females to a nest, where both birds call, the male sings, and the female often sits in the nest and probes at it. Unpaired males with a female nearby display and work at a nest much more vigorously than a paired male. The vigorous display of an unpaired male appears to strongly attract a female, even if she is already mated. On three occasions with color-banded birds, I saw a female that was already paired with one male fly into the territory of a vigorously displaying unmated male. Her original mate then flew after her and chased her back to their joint territory. In two cases, this occurred repeatedly throughout a day, the female's original mate continually retrieving her and generally ignoring the unpaired male. In one instance, the female had permanently switched the next morning and was paired with the previously unmated male.

The male builds almost all of the nest, but the female often helps as the nest nears completion. Males continuously work and display at nests until eggs are laid, and many build two or three nests. Nests are usually worked on serially, but occasionally males work on and display to two nests in the same day. Often an old nest is torn down for nest material. Less frequently, material is stolen from the nest of a neighboring Phainopepla or another species, usually Verdins (*Auriparus flaviceps*) or house finches.

BREEDING BEHAVIOR AND SUCCESS

BREEDING SUCCESS AND TIMING

Clutch records from 1897 to 1975 show that laying in the Colorado Desert occurs from the last week of February to the first week of April, with March 14 the mean date of clutch completion (fig. 8). Only two-egg clutches are known from the Colorado Desert (n = 54), with the exception of the 1973 season at Shaver's Wash (see below). In 19 nests I observed, 53 percent of the eggs produced fledged young.

Distinct annual and geographic variations exist in breeding success. In 1973 breeding Phainopeplas were common throughout the Colorado Desert. At Shaver's Wash, mean clutch size was the largest on record: 2.5 (n = 18). This included the only three-

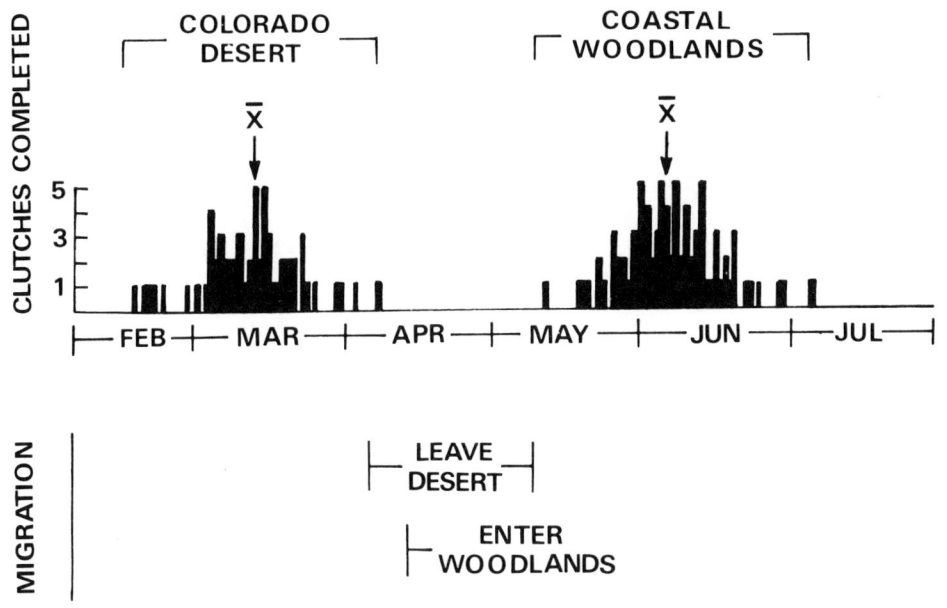

Fig. 8. Timing of egg-laying and migration in the Phainopepla's dual breeding ranges. Egg data are records of clutches completed on a particular date (see methods section). Mean date of clutch completion ($\bar{x}$) is shown for each habitat. The approximate timing of spring migration is shown for each area. The end of migration into the coastal woodlands is not distinct (see section on territoriality in the Santa Monica Mountains).

egg clutches known from the Colorado Desert. Yet along 5.2 km of Shaver's Wash, 29 territorial males and an unknown number of females produced 18 clutches, indicating that only 62 percent of the population bred.

No Phainopeplas bred at Milpitas Wash or surrounding areas in 1974, though a large population courted and built nests until migration. That year, Phainopeplas bred mainly in higher areas of the Colorado Desert, such as Shaver's Wash. Along 5.2 km of this wash, 27 territorial males and an unknown number of females produced 16 clutches, indicating that 59 percent of the population bred.

Fewer birds bred in 1975. No breeding birds were found in the following areas despite repeated visits throughout March and April: Milpitas Wash, Shaver's Wash, and three sites in Anza-Borrego Desert State Park: Yaqui Wells, Mountain Palm Springs, and Palm Spring. Shaver's Wash had no ripe mistletoe and thus no Phainopeplas. The other areas had large populations with males building nests, but no eggs were laid. A few breeding pairs were found along Coyote Creek in Anza-Borrego Desert State Park and at the Coyote Wash site. At Coyote Wash, 5 of 20 pairs laid eggs. No difference was detected between the territories of birds that bred and those that did not.

BREEDING BEHAVIOR

Males and females share incubation almost equally (tables 1, 2). For the two pairs for which time budgets were calculated, total attentiveness by both parents averaged 97.5

TABLE 2
A Comparison of Mean Time Budgets between Habitats
(behavioral categories)

	Flight	Perching	In transit	Eating fruit	Flycatching	At nest	Social activity
Nonbreeding males							
Percent of day spent in activity							
Santa Monica Mtns. (n = 3)	17.1	73.0	12.8	4.5	0.17	—	9.6
Colorado Desert (n = 4)	6.1	85.9	1.8	9.3	1.1	—	3.3
Ratio of values (SMM[1] : CD[2])	2.79	0.85	7.11	0.48	0.15	—	2.90
Statistically significant differences[3]	X	X	X	X			
Incubating males							
Percent of day spent in activity							
Santa Monica Mtns. (n = 3)	13.1	37.9	12.3	3.7	0.4	43.3	2.5
Colorado Desert (n = 2)	3.1	46.6	2.9	4.1	0.5	44.5	1.5
Ratio of values (SMM : CD)	4.23	0.81	4.24	0.90	0.80	0.97	1.72
Statistically significant differences	X		X				
Incubating females							
Percent of day spent in activity							
Santa Monica Mtns. (n = 3)	13.4	32.0	13.0	4.4	0.9	46.3	3.6
Colorado Desert (n = 2)	3.5	37.7	2.8	4.7	0.9	52.7	1.3
Ratio of values (SMM : CD)	3.83	0.85	4.64	0.94	1.00	0.88	2.77
Statistically significant differences	X	X	X			X	

TABLE 2 – Continued

	Flight	Perching	In transit	Eating fruit	Flycatching	At nest	Social activity
Males with nestlings							
Percent of day spent in activity							
Santa Monica Mtns. (n = 3)	24.3	62.3	13.0	3.9	8.7	7.7	4.4
Colorado Desert (n = 3)	14.2	72.5	2.7	4.3	8.4	8.4	2.4
Ratio of values (SMM : CD)	1.71	0.86	4.81	0.91	0.87	0.92	1.83
Statistically significant differences	X	X	X			X	
Females with nestlings							
Percent of day spent in activity							
Santa Monica Mtns. (n = 3)	24.3	52.1	16.2	4.8	11.3	8.4	7.3
Colorado Desert (n = 3)	15.9	68.7	2.4	4.7	13.5	9.2	1.7
Ratio of values (SMM : CD)	1.53	0.78	6.75	1.02	0.84	0.91	4.29
Statistically significant differences	X	X	X		X	X	X

[1] SMM = Santa Monica Mountains.
[2] CD = Colorado Desert.
[3] Student's t-test, $P < 0.05$.

percent of the active day or 98.8 percent of the 24-hour day. As Crouch (1939) reported, eggs hatch 14 days after the clutch is completed. Young are fed almost exclusively on insects for the first two to three days, after which fruit also becomes a major item in their diet. Insects remain a major food throughout the nestling period. This is reflected in the increased time spent flycatching by both males and females, which average, respectively, 19 and 25 times the values for the same sex while incubating (table 2). Both parents feed and brood the young, though females tend to do more (table 2). As Crouch (1939) reported, the young fledge when 19 to 20 days old. A few days after the young start to move around the territory, the male appears to lose interest in them and neither chases nor feeds them.

Assuming 14 days of incubation and a 20 day nestling period, clutch data indicate April 17 as the average date of fledging, though recently fledged young are generally seen throughout April and the first week of May. Much of this fledging occurs after nonbreeding Phainopeplas leave on migration. In 1973 all nonbreeding Phainopeplas left Shaver's Wash between April 6 and April 20. In 1974 they left Shaver's Wash and Milpitas Wash between April 10 and April 22. At Coyote Wash in 1975, the first large exodus occurred on March 30 and almost all nonbreeding birds had left by April 16. Large migratory flocks were seen in Anza-Borrego Desert State Park on March 31. As these birds leave, breeding Phainopeplas that remain expand their territories into newly vacated areas, as was quantified at Shaver's Wash in 1973 (fig. 9). When surveyed on April 6, there were 14 Phainopeplas in the area, including three pairs breeding. These birds' territories included 63 mistletoe clumps with at least 100 ripe berries per clump; the three breeding pairs averaged 9.3 mistletoe clumps per pair (fig. 9). By April 20 the nonbreeding birds had left, but the breeding pairs remained with young still in the nest and had expanded their territories into the newly vacated areas. The number of mistletoe clumps with at least 100 ripe berries had dropped 51 percent but due to their territorial expansions, these pairs experienced a decrease averaging only 14 percent (fig. 9). This characteristic territorial expansion may provide an important input of food for breeding birds at the time mistletoe fruit is becoming scarce. Most breeding Phainopeplas leave the Colorado Desert by the end of April, and only a few stragglers remain after the first week of May.

RIPARIAN WOODLAND SYSTEM

Resources

GENERAL PATTERNS OF FOOD USE

As in the Colorado Desert, fruit is the adult Phainopepla's major food in the Santa Monica Mountains. Berries eaten include those of elderberry (*Sambucus mexicana*), chaparral currant (*Ribes malvaceum*), summer holly (*Comarostaphylis diversifolia*), and buckthorn (*Rhamnus crocea*); as well as the buds of laurel sumac flowers. During April and May, Phainopeplas are usually found feeding on *Ribes*. They shift to *Rhamnus* when it starts fruiting at the end of May, as *Ribes* fruit is disappearing. *Rhamnus* usually remains the major fruit through the breeding period. Phainopeplas not feeding young averaged 4.2 percent of their active day feeding on fruit (table 1). As in the Colorado Desert, only adults feeding nestlings forage extensively on insects; flycatch-

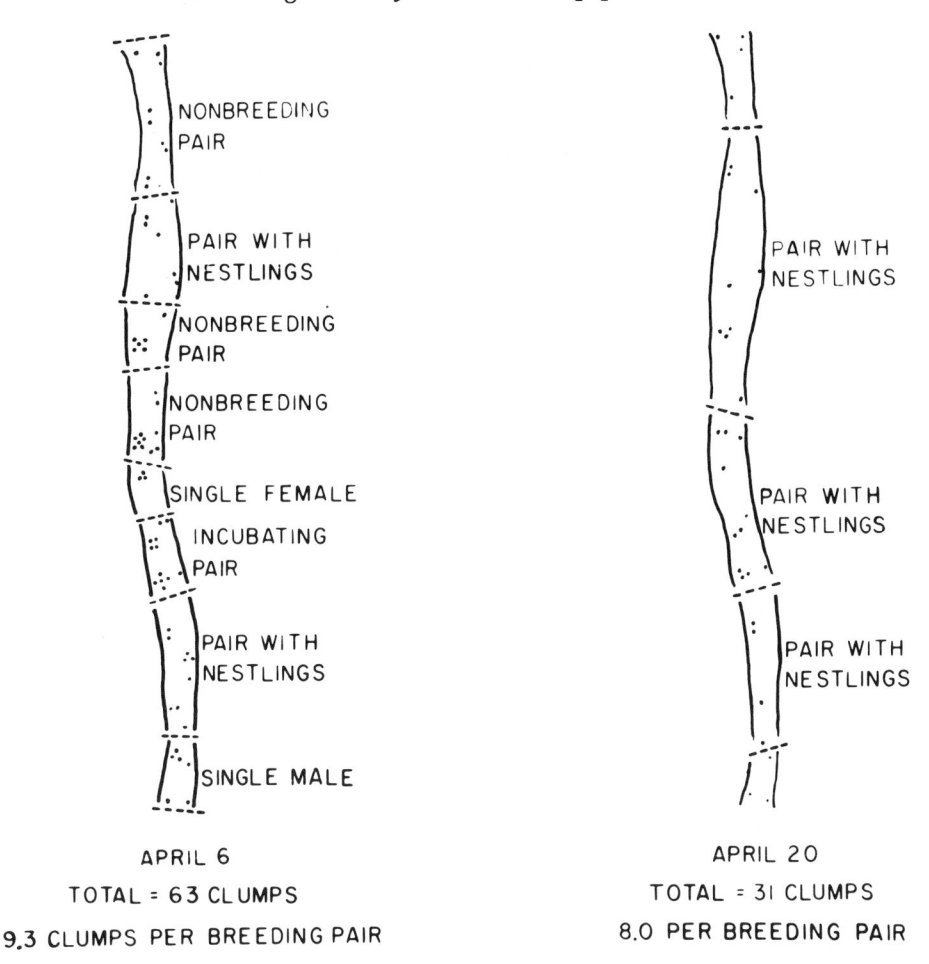

APRIL 6
TOTAL = 63 CLUMPS
9.3 CLUMPS PER BREEDING PAIR

APRIL 20
TOTAL = 31 CLUMPS
8.0 PER BREEDING PAIR

Fig. 9. Shaver's Wash in 1973, showing Phainopepla territories and desert mistletoe clumps bearing at least 100 ripe berries. Solid lines indicate wash borders, dashed lines indicate approximate territorial limits, and dots represent mistletoe clumps bearing at least 100 ripe berries.

ing a mean of 10 percent of their active day (table 2). This is 25 times the value (0.4%) for birds without young (table 1).

DISTRIBUTION AND ABUNDANCE OF FRUIT

As is typical of the fruiting shrubs that Phainopeplas feed upon in the Santa Monica Mountains, *Rhamnus* at Decker Canyon occurs on the chaparral-covered hillsides and is isolated from the riparian woodland in the canyon bottom (fig. 10). This is in sharp contrast with the spatial coincidence of fruit and relatively large trees in the Colorado Desert. Observations in 1972 and 1973 indicated that *Rhamnus* begins fruiting between late May and early June, peaks during the first half of July, and declines greatly by the first week of August. Censuses at Decker Canyon in 1974 confirmed this (fig. 11) and showed that fruit peaked while most Phainopeplas were feeding young. Individual

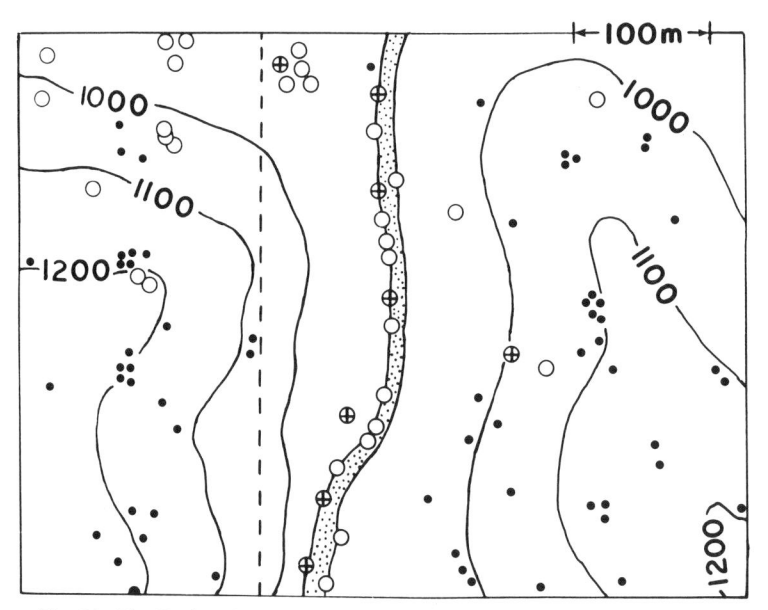

Fig. 10. The Decker Canyon site in 1974, showing *Rhamnus crocea* that fruited (small, solid circles), trees where Phainopeplas nested (circles with crosses), and other potential nest-trees (*Quercus* spp. and *Platanus racemosa*; large circles). Entire area mapped is 27 ha. Area left of dotted line is 9 ha area with 27 censused *Rhamnus*. Stippled area indicates stream bed. Road paralleling stream has been omitted. Elevations are in feet.

plants held abundant fruit for a short period (fig. 12), maintaining at least 50 percent of their maximum level a mean of 1.8 weeks (n = 27). Thus, the location of richly fruiting shrubs changed biweekly.

At Decker Canyon, *Rhamnus* was the only plant that produced fruit suitable for Phainopeplas during the breeding season. Within the 9.0 ha census area, 27 *Rhamnus* fruited (fig. 10). The standing crop of ripe berries peaked at 68,600 or a mean of 7,600 berries/ha (fig. 11). The mean dry weight of ripe *Rhamnus* berries is 49.5 mg. Averaging 4.78 cal/mg (S.D. = 0.28, n = 5), a berry contains 237 calories or 2.8 times the value for desert mistletoe. Thus, the peak standing crop had a caloric content of 1800 kcal/ha (fig. 11) or, respectively, 7.5 percent and 39 percent of the highest and lowest desert values measured. The actual energy available to a bird is a function of digestive efficiency, which has not been measured for Phainopeplas fed *Rhamnus* berries, but is probably similar to their 49 percent efficiency when fed mistletoe berries (Walsberg, 1975). However, even if digestive efficiency was 100 percent for birds fed *Rhamnus*, the qualitative relationship of the standing crops of available energy would not change.

INSECTS

The diurnal aerial-insect population was censused at Decker and Rustic canyons between June 5 and August 9, 1974 (fig. 13). Insects at Rustic Canyon averaged 604.9 mg/day (dry weight, n = 10), which is significantly higher than the 258.3 mg/day

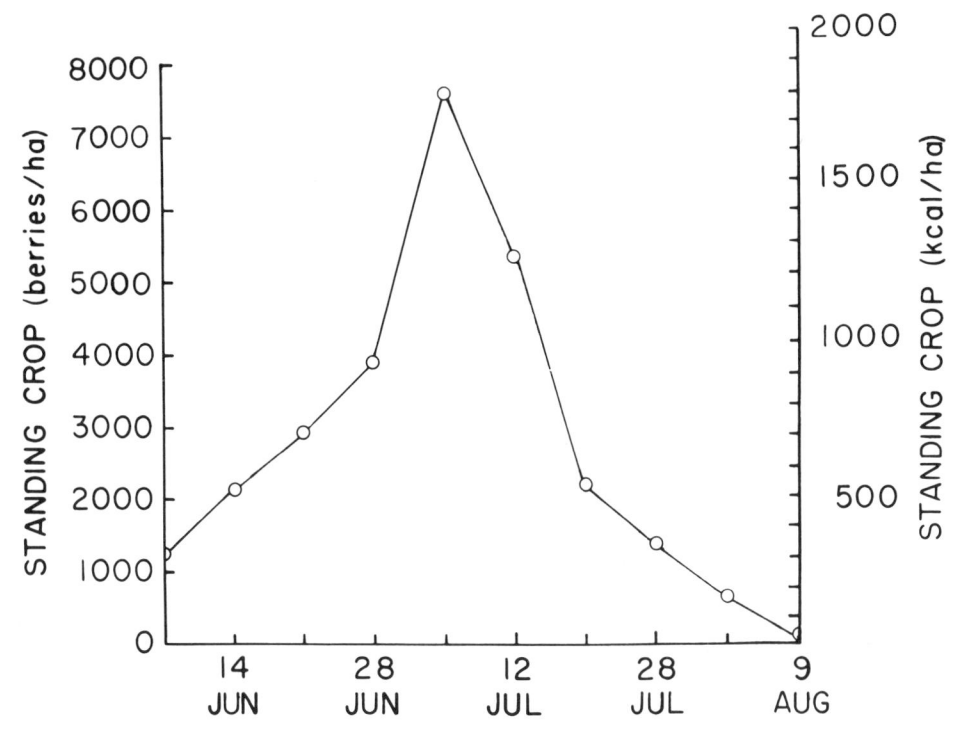

Fig. 11. Standing crop of ripe *Rhamnus crocea* berries during a 10-week period at Decker Canyon in 1974. Energy content calculated assuming 237 cal/berry (see text).

(dry weight, n = 5), which Decker Canyon averaged (Student's *t*-test, $P < 0.05$). Thus, Rustic and Decker canyons averaged, respectively, 32 and 13 times the peak insect abundance measured at Coyote Wash in 1975.

Territoriality

ARRIVAL AND ESTABLISHMENT

Phainopeplas were first seen in the Santa Monica Mountains on April 23 in 1972, April 12 in 1973, and April 19 in 1974. Numbers continued to be augmented greatly until mid-May. It is not obvious when an influx ends, since Phainopeplas frequently occupy an area for a number of days and then abandon it. Only when a pair has laid eggs can it be safely assumed that they will remain in an area. Thus, Phainopeplas do not occupy many sites until June, but this is apparently due to local movements and not migration. Males establish territories immediately upon arrival in a suitable area, while females defend only their mate's previously established territory.

CHARACTERISTICS OF TERRITORIES

Riparian woodland territories usually consist of a tree, typically a medium to large oak or sycamore, plus a small marginal area (figs. 14, 15). Territories in upper Rustic Can-

$0 < n \leq 0.5$ -----
$0.5 < n \leq 2.0$ ———
$2.0 < n \leq 3.5$ ▭
$3.5 < n$ ■

RIPE BERRIES PER LITER

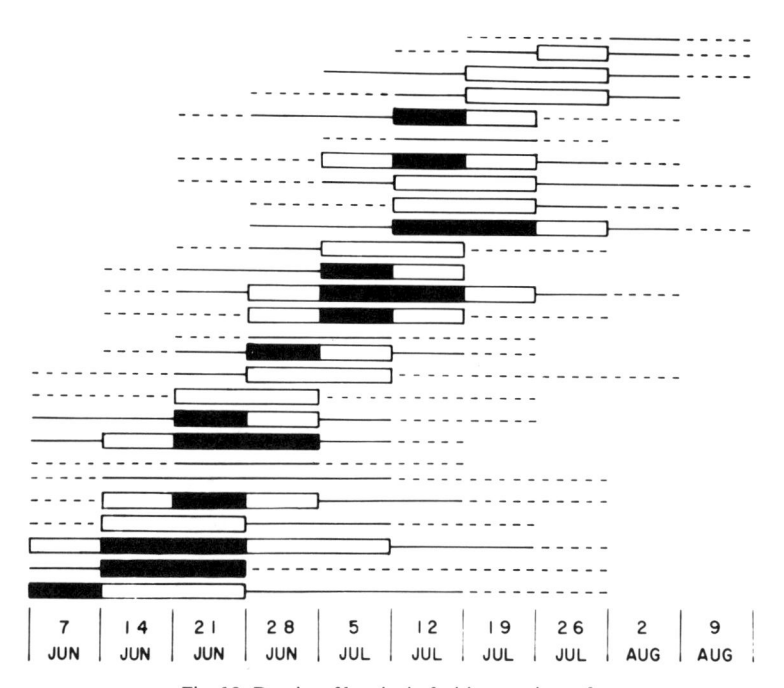

Fig. 12. Density of berries in fruiting portions of 27 *Rhamnus crocea* during a 10-week period at Decker Canyon in 1974.

yon were close together and averaged 0.023 ha (fig. 14). Territories in lower Rustic Canyon were spaced farther apart and averaged 0.040 ha (fig. 15). On two occasions, I observed two pairs nesting in the same tree, of which each defended a portion.

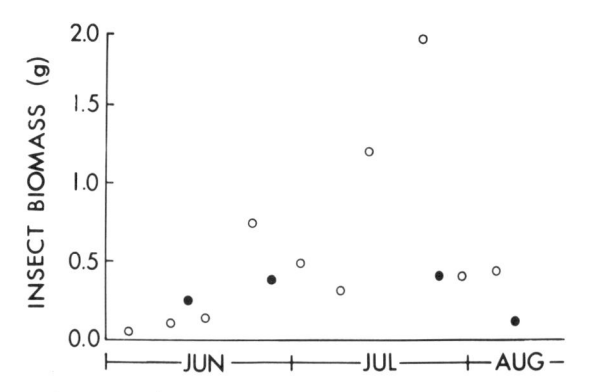

Fig. 13. Aerial insect abundance at Rustic Canyon (open circles) and Decker Canyon (solid circles) in 1974. Data are dry weights of insects caught per day in Malaise trap.

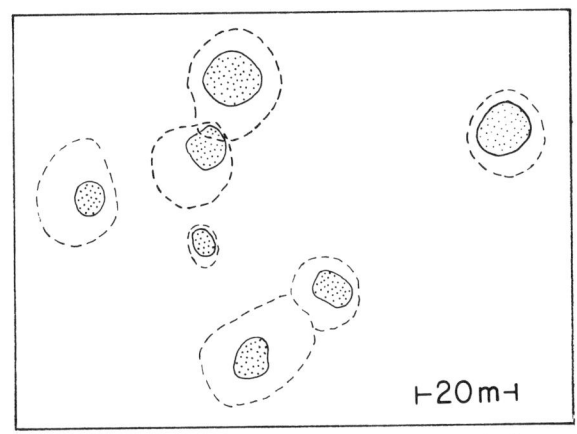

Fig. 14. Phainopepla territories in the upper Rustic Canyon colony in May 1974. Nest-trees are stippled and dashed lines indicate territorial boundaries.

DEFENSE AND ADVERTISEMENT

Territorial defense and advertisement differs only slightly from that seen in the Colorado Desert. Courtship flights are not used in defense (see section on courtship). Though advertisement is still primarily expressed by perching, it is much less conspicuous than in the desert; the background in the riparian woodlands is a chaparral-covered hillside and not the sharply contrasting desert sky.

Juveniles, apparently the product of the desert breeding, are seen occasionally during courtship and territorial establishment. Males defend their territories against these birds, but later, during July and August, both parents and juveniles drift into neighboring territories and are generally ignored.

COLONIALITY

Phainopeplas are not uniformly distributed throughout suitable areas in the Santa Monica Mountains, but tend to form loose colonies. Not all are colonial; solitary males or single pairs are not unusual. The location, date, and size of colonies seen were: Decker Canyon, June to August 1974, 8 territories (fig. 10); Decker Canyon, June to August 1973, 11 territories; upper Rustic Canyon, May 1974, 7 territories (fig. 14); lower Rustic Canyon, June to August 1974, 6 territories (fig. 15); Big Sycamore Canyon, June 1974, 4 territories; Santa Inez Canyon, June to August 1974, 6 territories; and Tuna Canyon, June 1974, 6 territories. In 1974, I searched suitable habitat two kilometers above and below the colonies in Decker, Tuna, Big Sycamore, and upper Rustic canyons. No other Phainopeplas were found, emphasizing the distinctness of these colonies.

Entire colonies apparently may shift location. When the upper Rustic Canyon colony was first found on May 20, males were building nests and establishing territories (fig. 14). Between May 25 and May 27, all but one pair left the area. Two males in this colony were banded and, between May 25 and May 28, they and four other

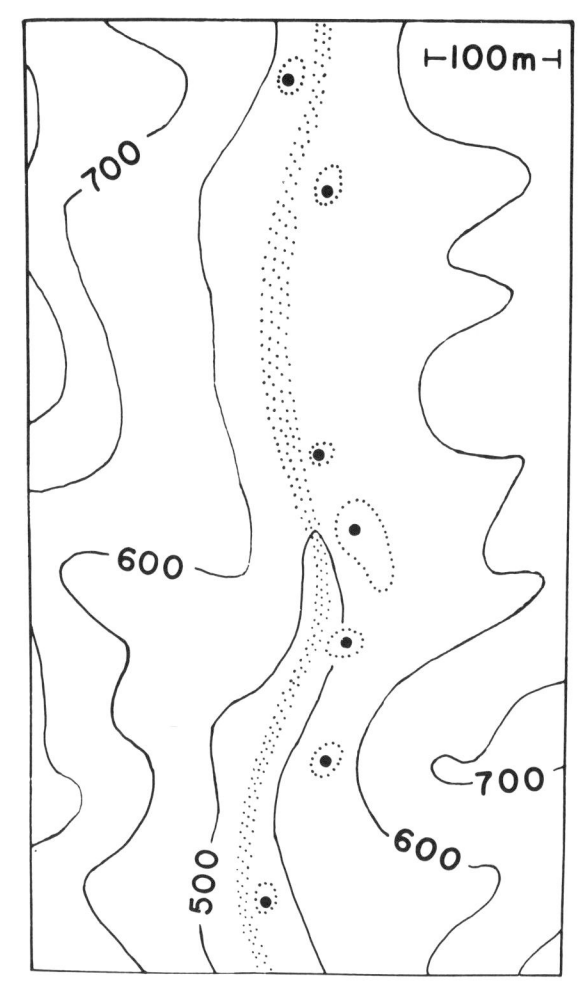

Fig. 15. Phainopepla territories in the lower Rustic Canyon colony in June, 1974. Nest-trees are shown by solid circles, and dotted lines indicate territorial boundaries. Stream bed is stippled. Contour lines are shown with elevations in feet.

males established territories about 1.2 km downstream in a site previously devoid of Phainopeplas (fig. 15). From this shift of banded birds, plus the close correspondence of numbers (seven males originally at the upper colony, then six at the lower and one at the upper), it appears that this colony moved en masse.

SOCIAL BEHAVIOR ASSOCIATED WITH FOOD

BEHAVIOR ASSOCIATED WITH FRUIT

Fruit was not defended successfully, except in one case wherein an isolated nest occurred very close (25 m) to a fruiting shrub. Although aggressive call-notes and displacements are common, it is not unusual to see a number of Phainopeplas feeding simultaneously at a shrub. Phainopeplas seem to prefer richly fruiting plants and usually feed at a discrete set of one to four shrubs during a day. At least occasionally, Phainopeplas locate a fruiting shrub by following another Phainopepla. This was seen in three instances with individuals having a known history of feeding sites. These birds were at their nest-tree when a second bird from the colony (not its mate) flew to, and fed at, a fruiting shrub in the hillside chaparral. The first bird immediately flew to this shrub also, though it had not previously fed within 100 m of the plant since it had started fruiting. The bird then fed at this new site for one to two weeks.

SOCIAL FLYCATCHING

Social flycatching was seen only in drier, more grassy sites with relatively few flying insects, such as Decker Canyon. There it occurred almost daily from the third week of June until the end of the nestling period in August. Only Phainopeplas feeding young were involved. A bout typically starts with a Phainopepla flying high (50-150 m) across a canyon or over a ridge, usually at dawn and rarely at dusk. The bird appears to intercept an insect swarm and starts flying in circular patterns and apparently flycatching. It is joined in this flycatching in 10 to 60 seconds by one to seven Phainopeplas flying in from the colony and surrounding hillsides. Cliff swallows (*Petrochilidon pyrrhonota*) joined the Phainopepla aggregation on three occasions, and twice, when it occurred low over a ridge, scrub jays (*Aphelocoma coerulescens*) and mockingbirds also joined in flycatching. Social flycatching lasts up to an hour, with birds intermittently leaving to feed young and returning. I observed two bouts from within 3 to 10 m as they occurred over a ridge I was on, and saw almost no aggression.

A social flycatching aggregation has an interesting spatial relationship to the lower edge of sunlight shining across a canyon at dawn. As the sun rises behind the eastern side of a canyon, the first direct sunlight crosses high over the canyon and strikes the opposite slope. All social flycatching occurs above the lower edge of this sunlight. As the morning progresses and the sunlight slants deeper into the canyon, the flycatching aggregation drops lower. When direct sunlight finally strikes the treetops, Phainopeplas switch to flycatching by short sallies out from their nest-trees.

This restriction to sunlit levels may be owing to the conspicuous visibility of even very small insects flying in direct sunlight. To human observers, these insects are often invisible while in the shade. Thus, the location of flycatching aggregations may be linked to that of the day's first visible insects.

COURTSHIP

Intensive courtship and nest construction occur in the Santa Monica Mountains as soon as the birds arrive in April and continue into July. Behavior resembles the last stage of desert courtship, with some differences. Courtship flights and assemblages are not seen. Courtship is centered in the nest-tree, though males are occasionally attracted from their territories by a female in the vicinity. Nest-display is the dominant feature of courtship and chasing is reduced compared to its incidence in the Colorado Desert. As in the desert, females often consort with a number of males before final pairing and there is no indication that birds retain mates from an earlier desert breeding.

BREEDING BEHAVIOR AND SUCCESS

Records from 1897 to 1974 of 86 clutches in the coastal oak and riparian woodlands of southern California indicate that the main laying period is the first half of June, with June 7 the mean date of clutch completion (fig. 8). At no time during this study was there any evidence of a significant nonbreeding population, either territorial or not. In the center of the 1974 breeding season, July 5 to July 9, 27 of 29 territories censused (93%) had breeding pairs. Two were held by unmated males.

Incubation behavior resembles that seen in the Colorado Desert, except that total attentiveness by both parents averaged 89.6 percent of the active day or 7.9 percent less than the corresponding desert value (table 2). This difference is statistically significant for females, but not for males (table 2). Incubating birds spent 1.7 (males) or 2.7 (females) times more time in social behavior than incubating desert birds (table 2). This increased social activity primarily involved mobbing scrub jays and was also seen in the nestling period, when riparian woodland birds spent 1.8 (males) or 4.3 (females) times more time involved in social behavior than desert birds (table 2). The behavior of adults with nestlings was otherwise similar to that seen in the desert, except that riparian woodland birds spent slightly (8%) less time at the nest (table 2).

Young fledge at the age of 20 days. In two instances juveniles left the nest and hopped through the nest-tree, then returned to the nest and did not leave finally until the next day. Fledged young generally remain in or near the nest-tree for one or two days, and then the entire family abandons the territory. Adults are often seen feeding young on surrounding hillsides and the young frequently perch directly in fruiting shrubs. Usually, this postfledging care is undertaken by the female; the male apparently loses interest in the female and the young once they abandon the territory. These females and their young do not respect territorial borders and commonly perch in another pair's nest-tree. The resident pair usually ignores them. Young apparently beg indiscriminately from females and may be fed by females other than their mother. In one instance, a tree contained a pair with nestlings, plus two other females and five fledglings. At least one of these fledglings begged from, and was fed by, each of the three females within one minute.

TIME AND ENERGY BUDGETS

TIME BUDGETS

Time budgets were calculated in both habitats for Phainopeplas of both sexes in three breeding statuses: nonbreeding, incubating, and caring for nestlings. No data were gathered on nonbreeding females in the Santa Monica Mountains, since these nonterritorial birds rarely remain in an area long enough to be banded and time-budgeted. I was able to time-budget only two pairs of incubating Phainopeplas in the Colorado Desert owing to the sparse breeding in 1974 and 1975. The following time budgets are of mated pairs: numbers 8 and 10, 9 and 11, 12 and 15, 13 and 16, 14 and 17, 21 and 24, 22 and 25, 23 and 26, 27 and 30, 28 and 31, 29 and 32 (table 1).

Time budgets expressed in behavioral categories are given in table 1, and are compared for birds of comparable breeding status in the Colorado Desert and in the Santa Monica Mountains in table 2. Table 3 presents the raw data for energy budget calcula-

TABLE 3
Time Budgets of Phainopeplas (activity level categories) and Temperature Data

No.[2]	Hours per day spent In flight	Hours per day spent In nonflight activity	Hours per day spent Inactive	T_m	$T_{m,50}$	$T_{m,25}$
1	1.09	8.41	14.50	13.2	13.9	14.1
2	0.20	9.40	14.40	15.4	16.8	17.3
3	0.15	9.67	14.18	15.6	17.1	17.6
4	1.08	10.41	12.51	18.9	19.8	20.1
5	0.19	9.07	14.74	13.3	14.4	15.0
6	0.61	10.01	13.38	12.7	13.5	13.5
7	0.54	9.66	13.80	15.7	17.1	17.5
8	0.37	11.30	12.33	12.7	12.9	13.8
9	0.35	11.59	12.06	18.7	18.9	20.2
10	0.45	11.33	12.22	12.7	12.7	14.4
11	0.38	11.51	12.11	18.7	19.3	19.6
12	1.82	11.09	11.09	16.4	18.0	18.0
13	2.16	11.00	10.84	17.7	19.4	19.6
14	1.59	11.41	11.00	17.6	19.5	19.8
15	2.11	10.81	11.08	17.0	18.7	19.2
16	1.86	11.27	10.87	18.4	19.9	20.3
17	2.24	10.82	10.94	18.6	20.0	20.1
18	2.48	12.12	9.40	18.7	19.5	19.9
19	2.37	12.08	9.55	20.1	20.9	21.3
20	2.62	12.21	9.17	20.0	21.5	21.7
21	1.87	12.58	9.55	18.4	19.4	20.0
22	1.87	12.61	9.52	19.5	20.4	20.6
23	1.99	12.60	9.41	20.0	21.2	21.8
24	2.03	12.49	9.48	20.2	21.4	21.4
25	2.01	12.72	9.27	19.3	20.0	20.5
26	1.80	12.58	9.62	18.7	20.1	20.2
27	3.80	10.81	9.39	19.7	20.2	20.7
28	3.38	11.30	9.32	19.5	20.1	20.6
29	4.20	10.32	9.48	19.3	20.0	20.5
30	3.16	11.51	9.33	19.2	19.3	19.6
31	3.89	10.47	9.64	19.7	20.6	20.7
32	3.52	11.01	9.47	20.2	20.9	21.0

Effective maintenance temperature (°C), calculated using three models[1]

[1] For a full description of temperature models, see methods section.

T_m: Black-bulb temperatures not used.

$T_{m,50}$: Black-bulb temperatures used for periods when the bird was in the sun at least 50% of the time.

$T_{m,25}$: Black-bulb temperatures used for periods when the bird was in the sun at least 25% of the time.

[2] See table 1 for site, date, status, and sex data.

tions. These are the same time budgets, but expressed as time spent in various activity levels. Associated temperature data are also given. Activity level budgets are compared for birds of comparable breeding status in the two habitats in table 4.

Birds in riparian woodland areas consistently spent more time in transit and social activity than did those in the desert. Both transit and social activity involve flight. This is reflected in the increased flight times for woodland birds, which average from 1.7 to

TABLE 4
A Comparison of Mean Time Budgets between Habitats
(activity level categories)

	Flight	Nonflight activity	Inactive
Nonbreeding males			
Hours per day in activity level			
Santa Monica Mtns. (n = 3)	2.49	12.14	9.37
Colorado Desert (n = 4)	0.63	9.47	13.89
Ratio of values (SMM[1] : CD[2])	3.95	1.28	0.67
Statistically significant differences[3]	X	X	X
Incubating males			
Hours per day at activity level			
Santa Monica Mtns. (n = 3)	1.91	12.60	9.49
Colorado Desert (n = 2)	0.36	11.45	12.20
Ratio of values (SMM : CD)	5.31	1.10	0.78
Statistically significant differences	X	X	X
Incubating females			
Hours per day at activity level			
Santa Monica Mtns. (n = 3)	1.95	12.60	9.46
Colorado Desert (n = 2)	0.42	11.42	12.17
Ratio of values (SMM : CD)	4.64	1.10	0.78
Statistically significant differences	X	X	X
Males with nestlings			
Hours per day at activity level			
Santa Monica Mtns. (n = 3)	3.79	10.81	9.40
Colorado Desert (n = 3)	1.86	11.17	10.98
Ratio of values (SMM : CD)	2.04	0.97	0.86
Statistically significant differences	X		X
Females with nestlings			
Hours per day at activity level			
Santa Monica Mtns. (n = 3)	3.52	11.00	9.48
Colorado Desert (n = 3)	2.07	10.97	10.96
Ratio of values (SMM : CD)	1.70	1.00	0.85
Statistically significant differences	X		X

[1] SMM = Santa Monica Mountains.
[2] CD = Colorado Desert.
[3] Student's t-test, $P < 0.05$.

5.3 times the desert values for comparable behavioral categories (table 4). The significantly decreased time spent inactive (asleep) and the tendency for increased nonflight activity in the Santa Monica Mountains reflects the increased daylength in midyear.

DAILY ENERGY EXPENDITURE (DEE)

DEE was calculated for each time budget using the three previously described models for the influence of direct solar radiation (table 5). Using the $T_{m,50}$ model, DEE is reduced an average of 1.9 percent below that calculated using the T_m model. The decrease is statistically significantly greater (Student's t-test, $P < 0.05$) for birds in the Colorado Desert ($\bar{x} = 2.3\%$) than for those in the Santa Monica Mountains ($\bar{x} = 1.5\%$; table 5). The $T_{m,25}$ model is a more liberal estimate of the effect of solar radiation; using it, energy expenditure is reduced only slightly more ($\bar{x} = 2.8\%$; table 5). Here also, the decrease is statistically significantly greater for desert values ($\bar{x} = 3.4\%$) than woodland values ($\bar{x} = 2.0\%$). DEE estimates do not appear very sensitive to these assumptions and for further analysis the intermediate $T_{m,50}$ values will be used.

DEE varied from 12.74 to 22.27 kcal/day (table 6), and changes in breeding status and habitat were associated with major differences in energy expenditure (table 7). Within each habitat, the energy expenditure of nonbreeding birds is roughly equal to that of incubating birds, while birds caring for nestlings expended about 20 to 30 percent more (table 7). For any particular breeding status, the DEE of birds in the Santa Monica Mountains was 19 to 28 percent higher than that in the Colorado Desert, and the differences are statistically significant (table 7).

Both absolutely and proportionately, the largest source of this increase in daily energy expenditure is the cost of flight. Riparian woodland birds expend 1.8 to 5.6 times as much energy for flight as do desert birds (table 7). Energy expenditure due to the additional cost of nonflight activity over maintenance levels is 2 percent lower for woodland birds than desert birds in some cases, and 10 to 28 percent higher for woodland birds than desert birds in other cases (table 7). Even where the cost of this nonflight activity makes its largest increase in woodland birds compared to desert birds (nonbreeding males), it is only 7 percent of the increased energy expenditure due to flight (table 7).

Recalculations of DEE using the daylength and temperature compensated model should roughly estimate the differences in energy expenditure due to resource patterns and social systems (tables 8, 9, 10). Values do not shift in any consistent direction compared to DEE estimated using the $T_{m,50}$ model and measured time budgets, and the 20 to 30 percent increased energy expenditure for woodland birds over Colorado Desert birds remains (table 10).

RELATION BETWEEN OXYGEN CONSUMPTION AND AMBIENT TEMPERATURE

Oxygen consumption was minimal between 30 and 43°C, where it averaged 2.44 ccO_2 g^{-1} hour^{-1} (fig. 16). This is 20 percent below that predicted by the Lasiewski-Dawson (1967) equation for a 26 g passerine (the mean weight of the experimental birds). Below 30°C, oxygen consumption was inversely related to ambient temperature (fig.

TABLE 5
The Effects of Three Temperature Models Upon Estimated Daily Energy Expenditure (DEE)[1]

No.[2]	T_m DEE[3]	$T_{m,50}$ DEE	% difference[4]	$T_{m,25}$ DEE	% difference
1	16.24	16.01	-1.4	15.94	-1.9
2	13.43	12.95	-3.6	12.78	-4.9
3	13.26	12.74	-3.9	12.57	-5.2
4	16.62	14.32	-2.1	14.22	-2.7
5	14.06	13.68	-2.7	13.48	-4.2
6	15.46	15.18	-1.8	15.18	-1.8
7	14.22	13.75	-3.4	13.61	-4.3
8	15.08	15.01	-0.5	14.70	-2.5
9	12.99	12.92	-0.5	12.47	-4.0
10	15.25	15.25	0.0	14.67	-3.8
11	13.06	12.86	-1.6	12.75	-2.4
12	17.43	16.91	-3.0	16.91	-3.0
13	17.86	17.31	-3.0	17.25	-3.4
14	16.05	15.88	-3.8	15.78	-4.4
15	17.92	17.38	-3.0	17.22	-4.0
16	16.91	16.43	-2.9	16.30	-3.6
17	17.74	17.30	-2.5	17.27	-2.7
18	18.52	18.27	-1.4	18.15	-2.0
19	17.80	17.55	-1.4	17.42	-2.1
20	18.49	18.02	-2.3	17.96	-2.9
21	17.14	16.81	-1.9	16.92	-3.0
22	16.78	16.49	-1.7	16.43	-2.1
23	16.93	16.54	-2.3	16.35	-3.4
24	16.95	16.57	-2.3	16.57	-2.3
25	17.22	17.00	-1.3	16.84	-2.2
26	16.86	16.41	-2.7	16.37	-2.9
27	21.41	21.26	-0.7	21.11	-1.4
28	20.46	20.28	-0.9	20.13	-1.6
29	22.48	22.27	-0.9	22.13	-1.5
30	20.02	19.99	-0.2	19.90	-0.7
31	21.59	21.33	-1.2	21.29	-1.4
32	20.57	20.36	-1.0	20.33	-1.2

$\bar{X} \pm$ S.D. (all birds): -1.9 ± 1.1 ; -2.8 ± 1.2
$\bar{X} \pm$ S.D. (desert birds): -2.3 ± 1.2 ; -3.4 ± 1.0
$\bar{X} \pm$ S.D. (woodland birds): -1.5 ± 0.8 ; -2.0 ± 0.8

[1] See text for description of temperature models.
[2] See table 1 for date, sex, breeding status, and habitat data.
[3] kcal/day.
[4] Calculated as: $\dfrac{\text{DEE} - (\text{DEE at } T_m)}{\text{DEE at } T_m} \times 100.$

TABLE 6
Daily Energy Budgets of Phainopeplas[1]

No.[2]	Maintenance[3]	Nonflight activity[4]	Flight[5]	Total daily expenditure (DEE)
1	11.47	1.28	3.26	16.01
2	10.91	1.43	0.61	12.95
3	10.83	1.47	0.44	12.74
4	9.51	1.58	3.23	14.32
5	11.75	1.38	0.55	13.68
6	11.85	1.52	1.81	15.18
7	10.65	1.47	1.63	13.75
8	12.18	1.72	1.12	15.01
9	10.12	1.76	1.03	12.92
10	12.21	1.72	1.33	15.25
11	9.97	1.75	1.14	12.86
12	9.78	1.68	5.44	16.91
13	9.19	1.67	6.46	17.31
14	9.40	1.73	4.75	15.88
15	9.43	1.64	6.31	17.38
16	9.15	1.72	5.56	16.43
17	8.97	1.64	6.70	17.30
18	9.02	1.84	7.41	18.27
19	8.63	1.84	7.08	17.55
20	8.34	1.85	7.83	18.02
21	9.31	1.91	5.59	16.81
22	8.99	1.91	5.59	16.49
23	8.68	1.91	5.95	16.54
24	8.60	1.90	6.07	16.57
25	9.06	1.93	6.01	17.00
26	9.11	1.91	5.38	16.41
27	8.26	1.64	11.36	21.26
28	8.47	1.72	10.10	20.28
29	8.16	1.56	12.55	22.27
30	8.80	1.75	9.45	19.99
31	8.11	1.59	11.63	21.33
32	8.17	1.67	10.52	20.36

Energy expenditure (kcal/day) due to

[1] Calculated using the $T_{m,50}$ temperature model (see methods section).

[2] See table 1 for habitat, date, breeding status, and sex data.

[3] Temperature-dependent maintenance expenditure for nonflight periods (see methods section).

[4] Does not include maintenance metabolism during periods of nonflight activity (see methods section).

[5] Total energy expenditure for periods spent in flight, including maintenance expenditure (see methods section).

TABLE 7
A Comparison of Mean Energy Budgets Between Habitats[1]

	Energy expended for			
	Maintenance[2]	Nonflight activity[3]	Flight[4]	Total daily expenditure (DEE)
Nonbreeding males				
Energy expenditure (kcal/day)				
Santa Monica Mtns. (n = 3)	8.66	1.84	7.44	17.95
Colorado Desert (n = 4)	10.68	1.44	1.88	14.00
Ratio of values (SMM[5] : CD)	0.81	1.28	3.96	1.28
Statistically significant differences[6]	X	X	X	X
Incubating males				
Energy expenditure (kcal/day)				
Santa Monica Mtns. (n = 3)	8.99	1.91	5.71	16.62
Colorado Desert (n = 2)	11.15	1.74	1.08	13.96
Ratio of values (SMM : CD)	0.80	1.10	5.29	1.19
Statistically significant differences		X	X	X
Incubating females				
Energy expenditure (kcal/day)				
Santa Monica Mtns. (n = 3)	8.93	1.91	5.82	16.66
Colorado Desert (n = 2)	11.09	1.73	1.23	14.05
Ratio of values (SMM : CD)	0.81	1.10	4.73	1.19
Statistically significant differences	X	X	X	X

TABLE 7 – Continued

	Energy expended for			Total daily expenditure (DEE)
	Maintenance	Nonflight activity	Flight	
Males with nestlings				
Energy expenditure (kcal/day)				
Santa Monica Mtns. (n = 3)	8.36	1.67	10.53	20.56
Colorado Desert (n = 3)	9.46	1.70	5.55	16.70
Ratio of values (SMM : CD)	0.88	0.98	1.90	1.23
Statistically significant differences	X		X	X
Females with nestlings				
Energy expenditure (kcal/day)				
Santa Monica Mtns. (n = 3)	8.30	1.64	11.34	21.27
Colorado Desert (n = 3)	9.18	1.67	6.19	17.04
Ratio of values (SMM : CD)	0.90	0.98	1.83	1.25
Statistically significant differences	X		X	X

[1] Calculated using the $T_{m,50}$ temperature model (see methods section).
[2] Temperature-dependent maintenance metabolism for nonflight periods only (see methods section).
[3] Does not include maintenance metabolism during periods of nonflight activity (see methods section).
[4] Total energy expenditure for time spent in flight, including maintenance expenditure (see methods section).
[5] SMM = Santa Monica Mountains; CD = Colorado Desert.
[6] Student's t-test, $P < 0.05$.

TABLE 8
Daylength-Compensated Time Budgets of Phainopeplas[1,2]

No.[3]	Hours per day spent In flight	Hours per day spent In nonflight activity
1	1.66	12.84
2	0.31	14.19
3	0.22	14.28
4	1.36	13.14
5	0.29	14.21
6	0.83	13.67
7	0.77	13.73
8	0.47	14.04
9	0.42	14.08
10	0.55	13.95
11	0.47	14.04
12	2.04	12.46
13	2.38	12.12
14	1.78	12.73
15	2.37	12.13
16	2.05	12.45
17	2.49	12.01
18	2.47	12.03
19	2.38	12.13
20	2.56	11.94
21	1.87	12.63
22	1.87	12.63
23	1.98	12.52
24	2.03	12.47
25	1.98	12.52
26	1.81	12.69
27	3.77	10.73
28	3.34	11.16
29	4.20	10.30
30	3.13	11.37
31	3.92	10.58
32	3.52	10.99

[1] Calculated assuming that birds were active 14.50 hours per day and they divided this period into their actually measured proportion of flight and nonflight activity (see methods section).
[2] Length of inactive period is 9.50 hours per day for all birds (= 24 - 14.50).
[3] See table 1 for site, date, breeding status, and sex data.

TABLE 9
Daily Energy Budgets of Phainopeplas
(daylength and temperature compensated)[1]

No.[2]	Maintenance[3]	Nonflight activity[4]	Flight[5]	Total daily expenditure (DEE)
1	9.20	1.95	4.97	16.12
2	9.76	2.15	0.92	12.84
3	9.80	2.17	0.65	12.62
4	9.33	1.99	4.08	15.40
5	9.77	2.16	0.87	12.79
6	9.55	2.08	2.48	14.10
7	9.57	2.08	2.31	13.96
8	9.70	2.13	1.39	13.22
9	9.72	2.14	1.25	13.10
10	9.66	2.12	1.64	13.42
11	9.70	2.13	1.39	13.22
12	9.05	1.89	6.11	17.04
13	8.91	1.84	7.11	17.86
14	9.16	1.93	5.31	16.39
15	8.91	1.84	7.08	17.84
16	9.04	1.89	6.13	17.06
17	8.86	1.82	7.44	18.13
18	8.87	1.83	7.37	18.07
19	8.91	1.84	7.10	17.85
20	8.83	1.81	7.66	18.31
21	9.12	1.92	5.60	16.63
22	9.12	1.92	5.60	16.63
23	9.07	1.90	5.91	16.89
24	9.05	1.89	6.06	17.00
25	9.07	1.90	5.92	16.89
26	9.14	1.93	5.42	16.48
27	8.33	1.63	11.28	21.24
28	8.51	1.69	9.98	20.19
29	8.16	1.56	12.55	22.27
30	8.60	1.73	9.34	19.67
31	8.27	1.61	11.72	21.60
32	8.44	1.67	10.51	20.61

[1] Calculated using data from table 9 (active day assumed equal to 14.50 hours) and assuming an average temperature of 20°C (see methods section).
[2] See table 1 for date, sex, breeding status, and habitat data.
[3] Temperature-dependent maintenance metabolism calculated for nonflight periods only (see methods section).
[4] Does not include maintenance metabolism during periods of nonflight activity (see methods section).
[5] Total energy expenditure for time spent in flight (includes maintenance metabolism; see methods section).

TABLE 10
A Comparison of Mean Energy Budgets Between Habitats
(daylength and temperature compensated)[1]

	Energy expended for			Total daily expenditure (DEE)
	Maintenance[2]	Nonflight activity[3]	Flight[4]	
Nonbreeding males				
Energy expenditure (kcal/day)				
Santa Monica Mtns. (n = 3)	8.87	1.83	7.38	18.06
Colorado Desert (n = 4)	9.52	2.07	2.66	14.24
Ratio of values (SMM[5] : CD)	0.93	0.88	2.77	1.27
Statistically significant differences[6]	X	X	X	X
Incubating males				
Energy expenditure (kcal/day)				
Santa Monica Mtns. (n = 3)	9.10	1.91	5.70	16.72
Colorado Desert (n = 2)	9.71	2.13	1.32	13.16
Ratio of values (SMM : CD)	0.94	0.90	4.32	1.27
Statistically significant differences	X	X	X	X
Incubating females				
Energy expenditure (kcal/day)				
Santa Monica Mtns. (n = 3)	9.09	1.91	5.80	16.79
Colorado Desert (n = 2)	9.68	2.12	1.12	13.32
Ratio of values (SMM : CD)	0.94	0.90	3.82	1.26
Statistically significant differences	X	X	X	X

TABLE 10 – Continued

	Energy expended for			Total daily expenditure (DEE)
	Maintenance	Nonflight activity	Flight	

Males with nestlings
Energy expenditure (kcal/day)
  Santa Monica Mtns. (n = 3)   8.44   1.67   10.52   20.63
  Colorado Desert (n = 3)   9.04   1.89   6.18   17.10
Ratio of values (SMM : CD)   0.93   0.88   1.70   1.21
Statistically significant differences   X   X   X   X

Females with nestlings
Energy expenditure (kcal/day)
  Santa Monica Mtns. (n = 3)   8.33   1.63   11.27   21.23
  Colorado Desert (n = 3)   8.94   1.85   6.86   17.66
Ratio of values (SMM : CD)   0.93   0.88   1.64   1.20
Statistically significant differences   X   X   X   X

[1] Calculated assuming a temperature of 20°C and an active day 14.5 hours long (see methods section).
[2] Temperature-dependent maintenance metabolism for nonflight periods only (see methods section).
[3] Does not include maintenance metabolism during periods of nonflight activity (see methods section).
[4] Total energy expenditure for time spent in flight, including maintenance metabolism (see methods section).
[5] SMM = Santa Monica Mountains; CD = Colorado Desert.
[6] Student's t-test, $P < 0.05$.

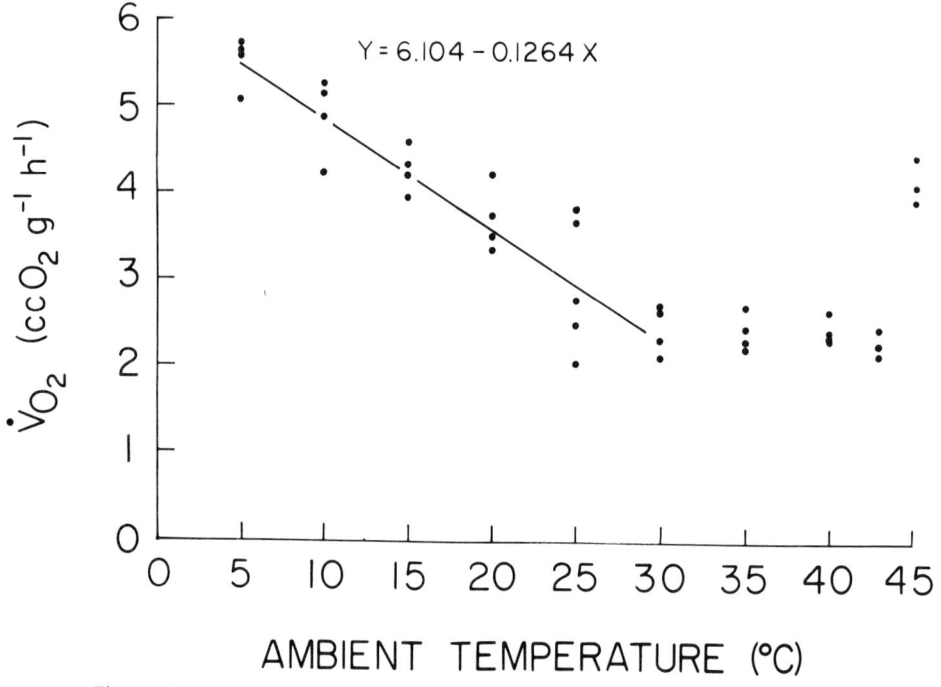

Fig. 16. The relation of oxygen consumption to ambient temperature in Phainopeplas. Line below 30°C was fitted by least-squares regression.

16). The least-squares regression line fitted to data collected below 30°C intersects the average thermal-neutral metabolism of 2.44 ccO_2 g^{-1} $hour^{-1}$ at 29°C (fig. 16).

DISCUSSION

ECOLOGY OF TERRITORIAL SYSTEMS

Defense of space by an animal entails expenditure of time and energy, as well as increased exposure to predation. When territoriality is exhibited, natural selection has presumably ensured that these costs are balanced by selective advantages to the individual. Variable expressions of territorial behavior should reflect differing balances of costs and benefits associated with resource defense (Brown, 1964). Phainopeplas exhibit a remarkable shift in territorial systems between the Colorado Desert and the Santa Monica Mountains, with individuals apparently using two contrasting systems in the same season (see section on the evolution of the annual cycle). Large feeding and nesting territories (type "A" of Nice, 1941) are defended along desert washes, while territories are much smaller and restricted to courtship and nesting in riparian woodlands (type "B"), where coloniality is also common. These differing territorial systems are correlated with sharply contrasting patterns of resources, and the data appear to be consistent with the general thesis that temporal and spatial patterns of food availability are the most important factors influencing territoriality (Brown and Orians, 1970).

COLORADO DESERT TERRITORIALITY

The selective advantages of type "A" territoriality (Nice, 1941) for the Phainopepla apparently center around the reduced and continuously decreasing fruit abundance during the breeding season (fig. 17). The critical nature of the fruit supply has been demonstrated by breeding failures of both individuals and large portions of populations due to insufficient fruit (see section on influences of resource availability upon breeding). An individual's defense of an often superabundant resource in midwinter, as seen at Milpitas Wash in 1974, may help to insure an adequate fruit supply during the breeding season. Because individual mistletoe clumps hold fruit for long periods, winter fruit abundance in a territory effectively predicts spring abundance. Note that in the three mistletoe-censused territories (fig. 6), the qualitative relationships of fruit abundance remained the same throughout the census period. The territory richest in fruit in January was also richest in April. The poorest territory in January contained no fruit by late March, when it was abandoned by the resident bird.

Type "A" territoriality probably also conserves available fruit by reducing energy expended in foraging. With a stable and evenly distributed food supply, nests spaced uniformly throughout a foraging area result in the shortest average foraging flight for each individual (Horn, 1968). The expected low energy expenditure due to flight was seen in the Colorado Desert (table 8), and may importantly conserve fruit while breeding by reducing consumption (see discussion section on energetics).

Type "A" territoriality is made possible by the simultaneous defensibility of fruit and nest-sites (fig. 17). Since mistletoe is a common parasite in the belt of trees and large shrubs used for nesting, an individual can defend both resources within a limited area. Defense of a feeding territory through the breeding season requires a stable food supply, which mistletoe provides. Seven weeks were required for the average mistletoe clump at Milpitas Wash to decrease 50 percent in fruit abundance, and even then clumps averaged 5,000 berries each.

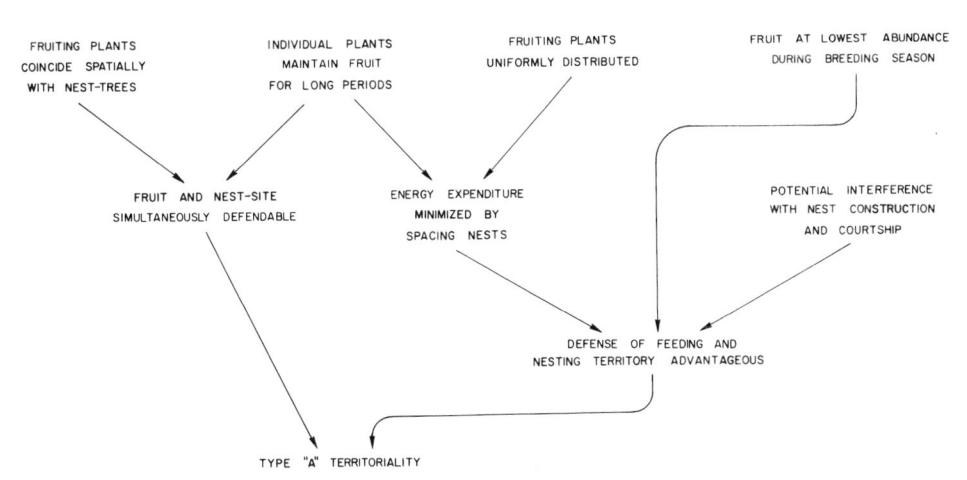

Fig. 17. The interaction of factors facilitating the Phainopepla's defense of a feeding and nesting territory (type "A" of Nice, 1941) in the Colorado Desert.

TERRITORIALITY IN RIPARIAN WOODLANDS

In contrast to the Colorado Desert, food and nest-sites are not simultaneously defendable in the Santa Monica Mountains (fig. 18). The location of abundant fruit is relatively unpredictable, since fruit is short-lived on a plant (fig. 12). In addition, fruiting shrubs are sparse, which militates against persistent defense of a combined feeding and nesting territory. Such a territory based upon plants that hold fruit for short periods is possible only if the plants are so dense that a limited, defensible area always contains at least one shrub with abundant fruit.

The spatial separation of fruit and nest-sites also acts against their simultaneous defense (fig. 18). The large trees preferred for nesting are concentrated in canyon bottoms, usually at least 100 m from fruiting shrubs. In both the Colorado Desert and the Santa Monica Mountains, Phainopeplas tend to nest in the largest trees or shrubs available, presumably to escape terrestrially-based predators. Suitability for nesting is obviously not a function of the plant's absolute size, since *Rhamnus* is often as large as desert nest-trees. However, nesting in *Rhamnus* would probably produce only a slight advantage in foraging, while nest predation may be reduced significantly by the use of the large oaks and sycamores. Potential nest-predators observed in shrubs, but not in the upper or middle layers of these large trees, include gopher snakes, *Pituophis melanoleucus*; Pacific rattlesnakes, *Crotalus viridis*; Beechey ground squirrels, *Citellus beecheyi*; and long-tailed weasels, *Mustela frenata*.

In riparian woodlands the Phainopepla's territory is essentially a tree used for nesting. This territoriality presumably reduces interference with courtship and nesting, of which examples were previously described (fig. 18). Indirect support for this function comes from the lack of defense late in the breeding season against females with young, which are not likely to interfere with the resident pair.

Type "B" territoriality and the loose coloniality seen in the Santa Monica Mountains are mutually reinforcing (fig. 18). The restriction of aggression to a very small ter-

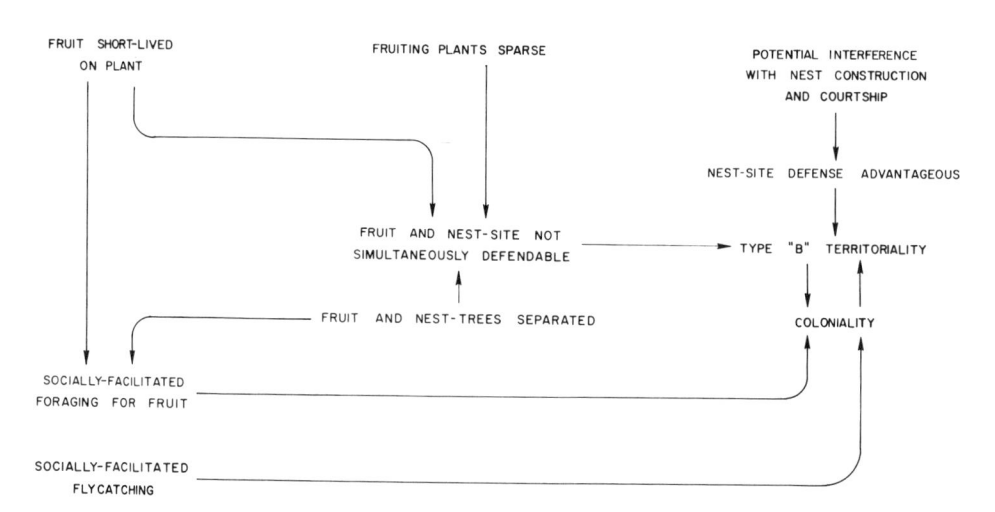

Fig. 18. The interaction of factors facilitating the defense of a restricted nesting territory (type "B" of Nice, 1941) and formation of colonies by the Phainopepla in the Santa Monica Mountains.

ritory facilitates coloniality, and the selective advantages of coloniality in turn favor reduction of territory size (see below).

COLONIALITY

Socially-facilitated foraging is apparently the major selective advantage of coloniality for Phainopeplas in the Santa Monica Mountains (fig. 18). Instances of one bird following another to a fruiting shrub were previously described and may occur frequently. I personally found that the easiest method of locating fruiting shrubs was to observe foraging flights from a colony. The hillside chaparral usually is seen easily from a colony and, because of their white wing patches and languid flight, Phainopeplas are easily followed. Additional evidence that Phainopeplas can, and do, observe the foraging flights of others is given by their immediate response to a bird flycatching high over a canyon at dawn, thus initiating a social flycatching bout.

The social flycatching seen in the Santa Monica Mountains is a clear example of socially-facilitated foraging (see section on social behavior associated with food). It is used only in twilight periods in relatively dry areas, when insect abundance is low and foraging difficult. Flycatching aggregations are presumably feeding upon insect swarms, which probably also occur in more lush canyons where social flycatching was not seen. However, insects in these canyons are probably so abundant that special foraging behavior is unnecessary.

COURTSHIP BEHAVIOR

Much of the Phainopepla's courtship behavior can be analyzed as adaptations to two major characteristics of its environment, which are most apparent in the Colorado Desert. They are the compressed and somewhat unpredictable timing of the breeding season, and the bird's occupation of visually-open habitats.

EMPHASIS UPON VISUAL DISPLAY

In relatively upon habitats, such as deserts and semiarid woodlands, visual conspicuousness may be a more effective signal over long distances than vocalizations, and be less expensive energetically. Therefore, it is not surprising that Phainopeplas emphasize visual display over vocalizations, as do a number of other birds inhabiting open areas. This is reflected in the bird's color and morphology, which makes Phainopeplas conspicuous at great distances. It is a common experience to perceive immediately a perching male more than 75 m away, yet be unaware of more brightly colored birds perching next to him, such as male house finches. This conspicuousness is aided by the Phainopepla's crest and long tail, which increase its apparent size, and by the white wing-patches when in flight. Conspicuous perching, aided by short call-notes, is the primary mode of territorial advertisement. Song is not used in this advertisement except when associated with courtship and is often inaudible to human observers more than 35 m away.

Courtship flights apparently function as long-distance signals indicating that a male in breeding condition is present. This display appears to be a ritualized form of the social flycatching seen in the coastal woodlands, to which it is strikingly similar in altitude, flight pattern, and crepuscular performance. The major differences between courtship flights and social flycatching are: (1) The snaps and flutters of flycatching are never seen in courtship flights; (2) the participants in courtship flights are as yet nonbreeding, while those flycatching are almost always feeding young; (3) no courtship and little aggression is seen in social flycatching; (4) social flycatching bouts usually last much longer than courtship flights. Ritualization of flycatching for courtship would not be surprising, considering the critical importance of insects for breeding (see section on influences of resource availability upon breeding).

ADAPTATIONS TO A COMPRESSED AND SEMIPREDICTABLE BREEDING SEASON

A compressed and only partially predictable breeding period is seen most clearly in the Colorado Desert, where Phainopeplas are apparently dependent upon the spring insect bloom, and also require a constant supply of desert mistletoe berries (see section on influences of resource availability upon breeding). Ripe mistletoe berries are always scarce by the beginning of May, while insects bloom in early to mid-March with the spring annuals. Thus, there is no more than a six- to eight-week period in which abundant insects coincide with sufficient ripe fruit. The Phainopepla requires five weeks from clutch completion until fledging. With the addition of postfledging care and migration, precisely timed breeding becomes critical. Conditions in the Santa Monica Mountains seem less severe, but it is probably advantageous for a bird to exploit a rich fruit supply as quickly as possible.

Adaptations facilitating a rapid response to favorable breeding conditions include nest-building by the unmated male and the early initiation of courtship. Intensive courtship began at the end of January at Milpitas Wash in 1974, six weeks prior to the mean clutch completion date in the Colorado Desert. Gould (pers. comm.) observed an even earlier initiation of courtship in mid-January of 1972 in the Avra Valley, north of Tucson, Arizona. This early courtship parallels that observed in a number of tropical and subtropical tyrannids that experience unpredictable and/or compressed breeding seasons (Wagner, 1941) and may allow Phainopeplas to prepare for the breeding period by reducing the male-female aggression that is residual from the establishment of winter territories. Characteristic of this early courtship is the female's association with a number of males. This may represent an inspection of the males and their territories, which provides the female with a wider choice of potential mates.

Data gathered at Milpitas Wash in 1974 showed that the initiation of nests by males coincided with, and was apparently stimulated by, the desert's spring bloom of insects. All males began nests in the seven-day period starting February 28, just as aerial-insect biomass was rapidly increasing (fig. 7). This phenomenon resembles the rain-stimulated nesting of other birds with unpredictable or compressed breeding periods, such as some Mexican tyrannids (Wagner, 1941) and many Australian birds (Immelmann, 1963). Nest construction by courting males should allow a rapid response to conditions sufficient for breeding. The most rapid completion of nests I have observed required four to five days and those begun as insect abundance first starts to rise should be complete when abundance reaches levels appropriate for breeding. It is apparently advantageous to maintain nests in good condition, since the time required for insect biomass to reach levels sufficient to support the birds' breeding is probably variable. This may be reflected in the female's inspection of the nest, accompanied by the male's nest-display and multiple-nest construction (see section on desert courtship). It is not obvious whether females base their choice of a mate to a large degree upon the condition of the nest, as is seen in Village Weaverbirds (*Ploceus cucullatus*; Collias and Collias, 1970), or upon other factors, such as the site of the nest or food availability.

DIFFERENCES IN COURTSHIP BETWEEN THE COLORADO DESERT AND THE SANTA MONICA MOUNTAINS

Conspicuous differences between courtship observed in desert washes and in riparian woodlands include the absence of courtship flights and courtship assemblages in the woodland and the immediate initiation of nests upon arrival in this habitat. This immediate start of nest construction may be interpreted in either of two ways. It is possibly a behavioral response to conditions potentially sufficient for breeding; for example, fruit is available and insects abundant. Alternatively, such conditions may not be necessary to release nesting behavior in the Santa Monica Mountains, and this nesting may represent a continuation of that nesting behavior begun in the desert. Sufficient data are not available to distinguish between these possibilities.

The absence of courtship flights and courtship assemblages in the Santa Monica Mountains is not well understood. It is possible that these displays are relevant only while both sexes defend separate territories. Both displays allow potential mates to become familiar with each other and to interact in situations where territorial aggression is minimized. During courtship assemblages, most birds involved are off their territories, except for the individual resident in the area where the courtship assemblage occurs. In courtship flights, potential mates meet at high altitudes over a territory, where defense is probably greatly reduced (see section on desert courtship). Such reduction of territorial aggression between potential mates is unnecessary in the Santa Monica Mountains, since intersex territoriality never occurs there.

An additional explanation for the absence of courtship flights in riparian woodlands may be the birds' frequent and conspicuous foraging flights to local hillsides. This makes unnecessary any special long-distance advertisement of a bird's presence. Social flycatching cannot similarly substitute for courtship flights, since it is characteristic only of adults feeding young.

INFLUENCES OF RESOURCE AVAILABILITY UPON BREEDING

INFLUENCES UPON THE TIMING AND SUCCESS OF BREEDING

Fruit is the major food of the adult Phainopepla, but the young are fed large amounts of insects as well. There are striking differences between the Colorado Desert and the Santa Monica Mountains in the timing of availability of these resources. Fruit in the Colorado Desert peaks in abundance in midwinter, while the abundance of flying insects peaks during the spring. This separation of the peak levels of these resources does not occur in the Santa Monica Mountains, where insects are abundant throughout the summer (Cody, 1974) and *Rhamnus* fruits most heavily in early July.

In the Colorado Desert, the Phainopepla's dependence upon insects for reproduction is demonstrated by its breeding in March and April. This is long past the period of peak fruit abundance, but is the only period in which abundant insects and sufficient fruit are simultaneously available. Insufficiency of either resource may prohibit breeding. Widespread breeding failures associated with low fruit production were seen in 1975. As ripe mistletoe at Coyote Wash disappeared in March, about 75 percent of the population left without breeding. Considering the generally poor mistletoe crop, it is doubtful if the birds that departed established territories elsewhere in the desert. Even in 1974, when berry production was high, one of the three mistletoe-censused territories totally lacked ripe fruit by the end of March.

For birds that do breed, the drop in fruit abundance during March and April may produce critically low levels before the young are able to migrate. As described previously, the emigration of nonbreeding birds may importantly reduce the breeding birds' competition for fruit (see section on Colorado Desert breeding behavior).

A different type of breeding failure, apparently created by low insect abundance, was seen at Milpitas Wash in 1974. Fruit was abundant, but the spring insect bloom reached levels only about one-half of that measured at Coyote Wash in 1975, when 20 percent of the population bred.

Influences on the timing of breeding in the Santa Monica Mountains are not obvious. Phainopeplas arrive in April and immediately start building nests, but few birds actually breed prior to the last third of May. This delay is associated with local shifts in areas occupied, and probably involves location of a final nest-site with sufficient fruit available. *Rhamnus* was the major fruit eaten by breeding birds in the areas I studied, though a few were found feeding mainly upon *Ribes* in May. In April and early May, *Ribes* accounts for most of the small amount of fruit in the chaparral that is suitable for Phainopeplas. It disappears at the end of May, when *Rhamnus* initiates fruiting. Birds laying eggs in May that depend upon *Ribes* may not be able to fledge young if sufficient *Rhamnus* does not fruit locally. The initiation of *Rhamnus* fruiting in early June coincides with the Phainopepla's peak laying period and the increased fruit abundance possibly stimulates laying. In any case, the two events are synchronized, so that the period of peak *Rhamnus* fruiting in July coincides with the period when Phainopepla young are near fledging or have just fledged. Thus, fruit is most abundant when consumption is high and inexperienced young must begin foraging.

INSECT ABUNDANCE AND CLUTCH SIZE

Phainopeplas lay clutches of either two or three eggs. Clutch size averages 2.46 in California's coastal mountains, significantly higher than the 2.00 value for the Colorado Desert in all years except 1973 (Student's t-test, $P < 0.05$). At Shaver's Wash in 1973, clutch size averaged 2.5. These differences are correlated with differences in insect abundance. This is supported by insect biomass data from the Santa Monica Mountains and the Colorado Desert, and from the circumstances surrounding the production of three-egg clutches at Shaver's Wash in 1973. Data gathered in 1974 and 1975 indicate that peak aerial-insect biomass coincides with the spring annual plant bloom. Associated with an exceptionally heavy fall and spring rainfall (U.S. Environmental Data Service, 1972, 1973), the flower bloom at Shaver's Wash in 1973 was the greatest I have seen in that portion of the Colorado Desert. Insect biomass probably was correspondingly high.

There are at least two nonexclusive explanations for an effect of insect biomass upon clutch size. The level of insect abundance during the egg-laying period is probably correlated with that during the nestling period. Since insect abundance should affect the optimal number of young which can be raised, selection will favor females that appropriately vary their clutch-size in response to differing levels of insect abundance. No data are available to evaluate this possibility.

It is also possible that the Phainopepla's egg-production is protein-limited and that abundant insects allow synthesis of a third egg. Protein content of Phainopepla eggs has not been measured, but can be estimated by assuming they contain 9.58 percent protein as do starling eggs (*Sturnus vulgaris;* Ricklefs, 1974). Fresh Phainopepla eggs average 2.78 g (Hanna, 1924), and thus would contain about 266 mg of protein. Phainopeplas lay one egg per day. Assuming four days are required for egg synthesis, the model of King (1973) predicts that peak demands of synthesis occur after the third day, when energy expenditure is equal to the synthesis of one egg per day. This estimate is based upon the growth rate of the ovarian follicle, and is probably proportional to protein demand. Thus, the maximum daily protein requirement for egg synthesis in a Phainopepla laying three eggs is equal to the protein content of one egg, after correction for inefficiency of synthesis. Assuming Ricklefs's (1974) 75 percent production efficiency of synthesis, the peak protein requirement is 355 mg/day ($=266/0.75$). The two obvious errors in this estimate are that egg synthesis may require more than four days, which would reduce peak demand (King, 1973), and that King's (1973) model prorated production of egg white over the entire period of egg synthesis. This should underestimate peak demand, since egg white is deposited in one day. These errors are in opposite directions, and the value of 355 mg/day is probably a reasonable estimate.

The amount of food a bird must ingest to acquire this protein can be estimated assuming the food to contain the necessary complement of essential amino acids and that borrowing from internal protein stores is insignificant. At most, such borrowing is probably very limited (Scott, 1973).

Assuming protein is 16 percent nitrogen (Jones, 1931), the 0.191 mg nitrogen per desert mistletoe berry (Walsberg, 1975) represents 1.19 mg of protein. This is 1.8 percent of fresh berry weight, or slightly higher than the 1.4 percent average for fruits analyzed by Leung (1968; in Jenkins, 1969). At the Phainopepla's 25 percent protein assimilation efficiency when fed mistletoe (Walsberg, 1975), a bird must ingest 1,183 berries per day to meet the peak protein requirement. Phainopeplas collected in the field typically have 11 to 16 berries in their digestive tract, though I have observed a maximum of 25. Marked berries pass through the digestive tract in 12 to 45 minutes ($\bar{x} = 29$ minutes, n = 17; Walsberg, 1975). Assuming the mean value of 29 minutes and that the digestive tract constantly contains 25 berries, then an average of 0.86 berries/minute can be processed. Processing 1,183 berries would require 23 hours. Thus assimilating 355 mg/day of protein during the Phainopepla's active period would be possible only if the assimilation rate was doubled, which seems unlikely.

In contrast, insects are probably a rich source of protein. Insects belonging to four orders were analyzed by Leung (1968; in Jenkins, 1969) and averaged 18 percent protein, or 10 times the value for desert mistletoe berries. Protein assimilation efficiency has not been measured for the Phainopepla or, apparently, for any passerine except the starling (Thompson and Grant, 1968). If the Phainopepla achieves the starling's

92 percent efficiency, then insects represent a protein source 37 times as concentrated as mistletoe. Insects probably also represent a protein source much richer than *Rhamnus*, if this fruit's protein content is similar to that of other fruits analyzed (Leung, 1968; in Jenkins, 1969).

USE AND EVOLUTION OF DUAL BREEDING RANGES

EVIDENCE FOR THE USE OF DUAL BREEDING RANGES

Breeding has been recorded throughout the Phainopepla's summer and winter ranges (see sources cited in fig. 1). The timing of residency and migration in the Colorado Desert and in the coastal woodlands of California leave little doubt that the same population is involved in both breeding periods in the two areas (fig. 8). Migratory flocks I observed flying west in San Gorgonio Pass, between the desert and the coast, in late April of 1972, 1973, and 1975 provide evidence of a direct migration of Phainopeplas from the Colorado Desert into the coastal woodlands. Such a migration is further supported by the sighting in late May of 1973 of a Phainopepla in the oak woodland at O'Neill Park, Orange County, in southern California, which had been marked as it was breeding in March at Shaver's Wash.

At present, sufficient data are lacking to state definitely whether individual Phainopeplas breed in both seasonal ranges in a single year. Certainly, ample time is available for individuals to do this. With 86 days between the mean clutch completion dates in the two seasonal ranges and a 34-day period from laying until fledging, 52 days are left for fledged young in the desert to become independent and for the adults to migrate into the coastal area and mate again (fig. 8). It is approximately 250 km from the center of California's portion of the Colorado Desert to the Santa Monica Mountains, and less to many other suitable coastal areas. Tucker's (1973) equation 62 predicts a minimum cost of transport speed of 7.92 m sec^{-1} for a 24 g Phainopepla, assuming a wingspan of 26.5 cm (the average of six wild-caught birds). At this speed, a Phainopepla can fly 250 km in 8.8 hours. Even assuming circuitous routes, sufficient time is obviously available for migration and remating.

It is possible, as Crouch (1943) suggested, that Phainopeplas that breed in the desert form a nonbreeding element in the coastal area. This hypothesis requires the existence of a large nonbreeding population in the coastal area which is nevertheless inconspicuous, for almost all adults observed in this area breed and all adults observed in both seasonal ranges court vigorously. I have been unable to detect any indication that such a nonbreeding population exists. Certainly, there is no obvious selective advantage for a Phainopepla that nested as early as March to refrain from a second breeding in June which could result in at least a doubling of the bird's annual reproductive output. Though the evidence is largely circumstantial, I feel that it strongly favors the hypothesis of a second nesting in the coastal area by individuals that earlier nested in the desert.

EVOLUTION OF DUAL BREEDING RANGES IN THE PHAINOPEPLA

The Phainopepla's use of dual breeding ranges may reflect the geological history of the plant communities it inhabits. During much of the Pliocene and pluvial periods of the

Pleistocene, plant communities characteristic of the Colorado Desert were more closely associated with chaparral and woodlands than at present (Axelrod, 1955, 1958). During the early Pliocene, woodlands and associated chaparral dominated the lowlands of the present Mojave and Sonoran deserts. Though no true desert occurred, arid subtropical scrub was scattered throughout the area. It was from this association that many contemporary Sonoran Desert wash plants were derived. Representatives of genera that currently support desert mistletoe are known from the subtropical scrub of this period, including *Cercidium, Olneya, Prosopis,* and *Acacia.* By mid-Pliocene, increasing aridity had restricted woodland and chaparral to higher elevations within the present desert area, which was generally dominated by grassland and mesquite-grassland. Desert scrub and arid subtropical scrub probably also occurred, though lowered winter temperatures and decreased summer rainfall had eliminated many subtropical arborescent elements from the present Mojave Desert. Pluvial periods of the Pleistocene brought a resurgence of chaparral and woodland in the present desert area, when biotic zones were lowered as much as 900 to 1200 m (Martin and Mehringer, 1965).

Thus, during much of the Pliocene and Pleistocene, Phainopeplas could have been very local or altitudinal migrants. In winter they may have resided at low elevations, along washes or in arid subtropical scrub, and fed upon the winter mistletoe crop. After breeding in the spring, they shifted up mountain slopes into woodland and chaparral areas, where they fed upon summer fruiting plants and laid a second clutch. As aridity increased and chaparral and woodlands were gradually excluded from the present desert area, Phainopeplas were forced to lengthen their migration routes into their present summer range. At present, migration is still on a very local or altitudinal basis in some areas. I have observed what is apparently altitudinal migration along the western edge of the Colorado Desert. Phainopeplas winter at the base of the peninsular ranges which border the desert, and summer in the woodlands at higher elevations. Scott (1885) described similar movements at the eastern edge of the Sonoran Desert. Very local movements are seen at the Big Morongo Wildlife Preserve in San Bernardino County, California. Here, a riparian woodland occurs in an otherwise desert area. Phainopeplas winter and breed in the mesquite patches which occur about 100 meters from this woodland. They vacate this mesquite in the spring. Phainopeplas then appear in the riparian woodland and breed. Whether or not the same individuals breed in both habitats is not known.

Energetics

ENERGETIC INSIGNIFICANCE OF INSOLATION

The estimated decrease in daily energy expenditure due to direct solar radiation averaged 1.9 percent using the $T_{m,50}$ model and 2.8 percent using the $T_{m,25}$ model (table 5). These values should be generally higher than those for other birds under the same conditions, since Phainopeplas have a relatively high lower critical temperature, associated with a basal metabolic rate which is 20 percent below that predicted by the Lasiewski-Dawson (1967) equation (see section on the relation of oxygen consumption to ambient temperature). Thus, insolation may reduce maintenance metabolism at higher temperatures.

The influence of solar radiation upon DEE can also be estimated using the most liberal effect of simulated solar radiation measured upon blackbirds. This is the 26 percent average reduction in standard metabolism of Cowbirds exposed to 0.9 cal cm^{-2} min^{-1} (Lustick, 1969). Time-budgeted Phainopeplas averaged 4.9 hours per day in direct sunlight (range, 1.5 - 7.8), or about 20 percent of a 24-hour day. A 26 percent decrease in maintenance metabolism for 20 percent of the day equals an average daily reduction of 5.2 percent (= 0.20 × 26%). Maintenance expenditure averages 59 percent of daily energy expenditure estimated using the T_m model (only shade air temperatures used). Thus, direct insolation probably reduces DEE maximally by 3.1 percent (= 0.59 × 5.2%).

This value is probably an overestimate, since periods of exposure to sunlight include most of the time in flight, when maintenance metabolism is not temperature-dependent. In addition, most of the Phainopepla's exposure to insolation occurs in the morning, less than three hours after sunrise, and in the hour prior to sunset. Recordings of solar radiation made at the Deep Canyon Desert Research Center (see methods section) show that radiation during these hours is always below the levels used by Hamilton and Heppner (1967) and Lustick (1969). Indeed, Phainopeplas avoid intense radiation even when sunning would be energetically advantageous, such as midday in January, when shade air temperatures are typically 10 to 15°C below their lower critical temperature.

It is apparent that reduction of energy expenditure by absorption of solar radiation is insignificant for Phainopeplas. However, their dark pigmentation may reduce energy expenditure by allowing effective use of conspicuous perching as an energetically inexpensive social signal.

COMPARISON OF DAILY ENERGY EXPENDITURE WITH THAT OF OTHER BIRDS

King (1974) has calculated the following least-squares regression from 18 measurements of DEE in 15 bird species:

$$\log \text{DEE} = \log 317.7 + 0.7052 \log W$$

where DEE is in kcal/day, and W is body weight in kilograms. This predicts a DEE of 22.90 kcal/day for a 24 g Phainopepla. For various behavioral categories, mean DEE ranges from 61 percent (incubating desert males) to 93 percent (woodland females with nestlings) of this value. This is equivalent to an average metabolic rate ranging from 2.1 to 3.2 times the measured basal metabolic rate (BMR), while the values summarized by King (1974) averaged 3.5 times estimated BMR. Indeed, the Phainopepla's DEE is so low that average values for nonbreeding and incubating birds in the Colorado Desert (table 7) fall under the parallel, but lower, regression line for DEE in rodents (King, 1974).

ECOLOGICAL SIGNIFICANCE OF DIFFERENCES IN ENERGY EXPENDITURE

Energy may be limiting to an animal associated with a restricted supply in the environment, a restricted rate of acquisition and assimilation, or limitations at the tissue level upon power available. It is doubtful if Phainopeplas approach this third possible limit, which would occur when an animal expends the maximum power available from various tissues, since the Phainopepla's energy expenditure is lower than that of other similar sized passerines (King, 1974; see above).

However, the Phainopepla's distribution and reproduction may be limited in both the Santa Monica Mountains and the Colorado Desert due to restricted availability of energy in the environment. In the Santa Monica Mountains, Phainopeplas may be excluded from otherwise marginally suitable areas due to their relatively high energy requirements (20-30% above desert levels) and the area's generally low fruit abundance. However, it is likely that the increased fruit consumption associated with these higher energy requirements has only a minor effect upon fruit availability. Phainopeplas do not defend food in the Santa Monica Mountains and, in sharp contrast to desert mistletoe, chaparral fruit is eaten by a great number and variety of animals. I observed the following species feeding upon *Rhamnus* berries: Audubon cottontails (*Sylvilagus auduboni*), brush rabbits (*Sylvilagus bachmani*), Beechey ground squirrels, coyotes (*Canis latrans*), raccoons (*Procyon lotor*), mule deer (*Odocoileus hemionus*), California quail (*Lophortyx californicus*), California thrashers (*Toxostoma redivivum*), mockingbirds, and house finches. These animals apparently consume a large portion of the berry crop, which greatly reduces the relative impact of the Phainopepla's consumption. Thus, the 20 to 30 percent additional energy intake required in the Santa Monica Mountains probably has a comparatively insignificant effect upon food availability.

In contrast, the relatively low energy expenditure in the Colorado Desert may importantly influence food availability. There are fewer frugivorous species in the Colorado Desert than in the Santa Monica Mountains, and none feed upon mistletoe as extensively as the Phainopepla. This is probably due to the Phainopepla's defense of this fruit and its digestive specializations for a mistletoe diet. The absence of these specializations may preclude other species from making effective use of this fruit (Walsberg, 1975). Thus, a Phainopepla is responsible for most of the berries consumed in its territory, and the reduced berry consumption associated with relatively low energy requirements may significantly conserve food. Since mistletoe fruit may become limiting during the breeding season, this conservation of food may be important in allowing residency and breeding in marginal areas or years of poor fruit production.

It is possible that Phainopeplas may be limited by their maximum rate of energy assimilation. Assuming that birds maintain energy balance, average hourly assimilation rates can be estimated by dividing a bird's daily energy expenditure by its time available for feeding (the length of its active day). The highest average assimilation rates required occur when daylength and daily energy expenditure are both either highest (adults feeding nestlings in the Santa Monica Mountains) or lowest (nonbreeding birds in the Colorado Desert; fig. 19). The required assimilation rates for these two categories are very similar in spite of the 48 percent higher average DEE of adults with nestlings in the Santa Monica Mountains. The highest required assimilation rate (1.69 kcal/hour) is that of a nonbreeding desert male on January 23 (budget 1, tables 1, 3, 6). This bird spent 11.5 percent of its active day in flight, which was associated with vigorous territorial defense against a persistently intruding Phainopepla. Previously, I estimated the maximum rate of berry processing through the Phainopepla's digestive tract as 0.86 berries/minute (see section of influences of resource availability upon breeding). At the Phainopepla's 49 percent digestive efficiency (Walsberg, 1975), this

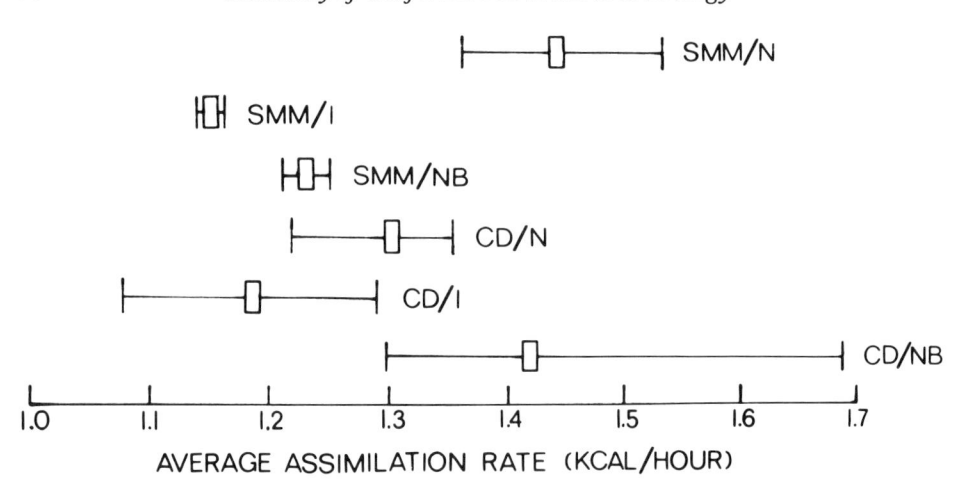

Fig. 19. Average assimilation rates required to maintain energy balance for Phainopeplas in various behavioral categories in the Santa Monica Mountains and the Colorado Desert. Data for males and females are combined, since mean values in various behavioral categories differ less than 5 percent between sexes. Values calculated as DEE/hours active; data from tables 3, 4, 7, and 8. Mean values represented by rectangles, ranges shown by vertical bars. SMM, Santa Monica Mountains; CD, Colorado Desert; NB, nonbreeding; I, incubating; N, feeding nestlings.

is equivalent to assimilating 2.12 kcal/hour. The required assimilation rate of the male on January 23 is 80 percent of this value. The maximum assimilation rate is probably approached most closely in December, when daylength is shortest and most territories are first established, requiring vigorous defense over a number of days. At this time, the Phainopepla's low maintenance metabolism and its extensive use of conspicuous perching as a means of inexpensive advertisement may importantly facilitate maintenance of the bird's energy balance and, therefore, its establishment of a territory.

In the Colorado Desert winter, foraging flights represent a minor expense of time and energy (tables 1, 2, 6, 7). In contrast, such foraging flights in the Santa Monica Mountains are a major source of energy expenditure. Theoretically, there is a hyperbolic relationship between required foraging time and the rate at which food is harvested during foraging (Epting, 1975; Gill and Wolf, 1975). The curve relating these two factors is such that, in the high range of foraging times, slight increases in harvesting rate may greatly decrease time required for foraging. This may be critical for birds feeding nestlings in the Santa Monica Mountains, which must forage for both themselves and their young, and have high required assimilation rates (fig. 19). Unfortunately, data are not available to estimate the maximum assimilation rate of Phainopeplas feeding on chaparral fruit. However, if adults feeding nestlings approach this unknown limit, then the reduction of energetically expensive foraging flights by increased efficiency in foraging becomes highly important. This gives added significance to the extensive use of socially-facilitated foraging by these adults.

ENERGY EXPENDITURE DURING BREEDING

The energy expended during breeding after the eggs are laid (E_b) may be estimated as:

$$E_b = (14 \times E_i) + (20 \times E_n)$$

where E_i is the mean daily energy expenditure of an incubating Phainopepla (table 7), the incubation period is 14 days, E_n is the mean daily energy expenditure of a Phainopepla with nestlings (table 7), and the nestling period is 20 days. This model estimates that 22 percent more energy is expended in the Santa Monica Mountains during breeding than in the Colorado Desert (table 11).

This estimate assumes uniform costs throughout the nestling period. Though energy expenditure probably changes significantly during this stage, measured values may approach an average, since adults were timed when their young reached approximately 50-60 percent adult weight (see methods section). This model does not include a number of other costs related to breeding. The cost of nest-construction cannot be isolated owing to its extensive use in courtship and the building of multiple nests. The energetic costs of gonadal growth and sperm production are ignored; they are probably insignificant (Ricklefs, 1974). Apparently also insignificant compared to energy expended during the incubation and nestling periods is the energetic cost of synthesizing eggs. A Phainopepla egg contains 2.86 kilocalories, assuming it weighs 2.78 g (Hanna, 1924) and contains the 1.03 kcal/g measured for starling eggs (Ricklefs, 1974). Assuming a 75 percent production efficiency (Ricklefs, 1974), the synthesis of one egg requires 3.81 kcal. The cost of producing the maximum clutch of three in the Santa Monica Mountains is about 11.4 kcal, or 1.7 percent of the 658 kcal the female expends during the incubation and nestling period (table 11).

It is interesting that mean clutch size in the coastal woodlands (2.46) is increased over that in the Colorado Desert (2.00) in about the same proportion as energy expenditure—23 percent. These parallel increases appear to be essentially independent. Clutch size is correlated with insect abundance (see section on influences of resource availability upon breeding), while differences in energy expenditure are primarily a function of the length of foraging flights to fruit. No significant effect of clutch size upon energy expenditure was seen in the three woodland pairs studied. The pair that raised three young (budgets 29 and 32, table 6) expended 42.63 kcal/pair/day during the nestling period, while the two pairs that raised two young (budgets 27, 28, 30, 31;

TABLE 11
Energy Expended During Breeding[1,2]

	Habitat		
	Santa Monica Mountains	Colorado Desert	Ratio[3] (SMM : CD)
Males	645	529	1.22
Females	658	538	1.22
Total per pair	1303	1067	1.22

[1] Kilocalories.
[2] See text for methods and discussion.
[3] SMM = Santa Monica Mountains; CD = Colorado Desert.

table 6) averaged 41.43 kcal/pair/day, or about 3 percent less. One may also note that the proportional difference in the energy expenditure of breeding birds between the two habitats is similar to that of nonbreeding males, where there is no effect of clutch size (table 7). Thus, energy expenditure during breeding is about 22 percent greater in riparian woodlands, but this is balanced by an increased clutch-size, so that average energy expended per breeding pair is remarkably similar for each egg laid; 530 kcal/egg in the Santa Monica Mountains and 534 kcal/egg in the Colorado Desert. This < 1 percent average difference should also apply to the energy expended per young fledged, since the proportion of eggs that produce fledged young does not differ significantly between the two habitats (Student's t-test, $P > 0.30$). A complicating factor is that the value for woodland birds is a population average, composed of pairs that raise two young and expend 22 percent more per young than desert pairs, and pairs that raise three young and expend 19 percent less per young than desert pairs. Available data are not sufficient to evaluate whether the convergence of average values between the two habitats is a fortuitous event, or if it is a reflection of an evolutionarily favored cost : benefit ratio in breeding.

The 3 percent increase in energy expenditure associated with the woodland pair caring for a third nestling is surprisingly small. This pair spent an average of 20 percent more time feeding on fruit and 31 percent more time flycatching than did woodland pairs with two nestlings. However, this was associated with only a 15 percent average increase in time spent in flight. These data may not be representative of all pairs raising three nestlings, since they are based upon only one such pair and time-budgets were calculated for each individual on only one day. It does, however, indicate that differing brood sizes may not necessarily entail large differences in parental energy expenditure.

REFERENCES

American Ornithologists' Union. 1957. Checklist of North American birds, 5th edition. Baltimore, Md.: Lord Baltimore Press. 691 p.

Axelrod, D. I. 1955. Evolution of desert vegetation in western North America. Carnegie Inst. of Washington Publ. 590:215-306.

———. 1958. Evolution of the Madro-Tertiary geoflora. Bot. Rev. 24:434-509.

Bailey, F. M. 1928. Birds of New Mexico. New Mexico Dept. of Game and Fish, Washington, D.C. 807 p.

Berger, M., and J. S. Hart. 1972. Die Atmung beim Kolibri *Amazilia fimbriata* wahrend des Schwirrfluges bei verschiedenen Umgebunstemperaturen. J. Comp. Physiol. 81:363-380.

Brown, J. L. 1964. The evolution of diversity in avian territorial systems. Wilson Bulletin 76:160-169.

Brown, J. L., and G. H. Orians. 1970. Spacing patterns in mobile animals. Ann. Rev. Ecol. Syst. 1:239-262.

Calder, W. A. 1964. Gaseous metabolism and water relations of the Zebra Finch, *Taenopygia castanotis*. Physiol. Zool. 37:400-413.

Cody, M. L. 1974. Competition and the structure of bird communities. Mongr. Pop. Biol. no. 7. Princeton, N. J.: Princeton University Press.

Collias, N. E., and E. C. Collias. 1970. Some experimental studies on the breeding biology of the Village Weaver, *Ploceus (Textor) cucullatus* (Muller). Ostrich Sup. 8:169-177.

Cowles, R. B. 1936. The relation of birds to seed dispersal of desert mistletoe. Madroño 3:352-356.

———. 1972. Mesquite and mistletoe. Pacific Discovery 25:19-25.

Crouch, J. E. 1939. Studies on the life history of *Phainopepla nitens lepida* Van Tyne and correlated changes in the testes. Ph.D. dissertation, University of Southern California.

———. 1943. Distribution and habitat relations of the Phainopepla. Auk 60:319-332.

Epting, R. J. 1975. Power input for hovering flight in hummingbirds and its effect on the time and energy budgets of foraging. Ph.D. dissertation, University of California, Los Angeles.

Gessaman, J. A. 1973. Methods of estimating the energy cost of free existence, pp. 3-31. *In* J. A. Gessaman, ed., Ecological energetics of homeotherms. A view compatible with energy modeling. Monograph series, vol. 20. Logan: Utah State University Press.

Gill, F. B. and L. L. Wolf. 1975. Economics of feeding territoriality in the Golden-winged Sunbird. Ecology 56:333-345.

Greenway, J. C. 1960. Family Bombycillidae, pp. 369-373. *In* E. Mayr and J. C. Greenway, ed., Checklist of birds of the world, vol. 9. Mus. Comp. Zool., Cambridge, Mass.

Grinnell, J., and A. H. Miller. 1944. The distribution of birds in California. Pac. Coast Avif. 27. 608 p.

Grinnell, J. and T. Storer. 1924. Animal life in the Yosemite. Berkeley: University of California Press.

Hamilton, W. J., III, and F. H. Heppner. 1967. Radiant solar energy and the function of black homeotherm pigmentation: an hypothesis. Science 155:196-197.

Hanna, W. C. 1924. Weights of about three thousand eggs. Condor 26:146-153.

Hart, J. S. and M. Berger. 1972. Energetics, water economy, and temperature regulation during flight. Proc. 15th Int. Ornith. Congr.:189-199.

Horn, H. S. 1968. The adaptive significance of colonial nesting in the Brewer's Blackbird (*Euphagus cyanocephalus*). Ecology 49:682-694.

Immelmann, K. 1963. Drought adaptations in Australian desert birds. Proc. 13th Int. Ornith. Congr.: 649-657.

Jaeger, E. C. 1957. The North American deserts. Stanford, Calif.: Stanford University Press. 308 p.

Jenkins, R. 1969. Ecology of three species of saltators with special reference to their frugivorous diet. Ph.D. dissertation, Harvard University. 318 p.

Johnson, D. H., M. D. Bryant, and A. H. Miller. 1948. Vertebrates of the Providence Mountains. Univ. Calif. Publ. Zool. 48. 375 p.

Jones, D. B. 1931. Factors for converting percentages of nitrogen in foods into percentages of protein. U.S. Dept. Agric. Circ. 183.

Kendeigh, S. C. 1973. pp. 111-117. *In* D. S. Farner, ed., Breeding biology of birds. Nat. Acad. Sci., Washington, D.C.

Kincer, J. B., 1941. Climate and weather data for the United States, pp. 685-699. *In* Climate and man (U.S. Dept. of Agric. yearbook of agriculture). U.S. Govt. Printing Office, Washington, D.C.

King, J. R. 1973. Energetics of reproduction in birds, pp. 78-117. *In* D. S. Farner, ed., Breeding biology of birds. Nat. Acad. Sci., Washington, D.C.

———. 1974. Seasonal allocation of time and energy resources in birds, pp. 4-70. *In* R. A. Paynter, ed., Avian energetics. Nuttall Ornith. Club, Harvard Univ., Cambridge, Mass.

Kontogiannis, J. E. 1968. Effect of temperature and exericse on energy intake and body weight of the White-throated Sparrow (*Zonotrichia albicollis*). Physiol. Zool. 41:54-64.

Lasiewski, R. C. 1963. Oxygen consumption of torpid, resting, active, and flying hummingbirds. Physiol. Zool. 36:122-140.

Lasiewski, R. C., and W. R. Dawson. 1967. A re-examination of the relation between standard metabolism and body weight in birds. Condor 69:13-23.

Leung, W. T. W. 1968. Food composition tables for use in Africa. FAO-HEW, Public Health Service, Bethesda, Md.

Lustick, S. 1969. Bird energetics: effects of artificial radiation. Science 163:387-390.

Martin, P. S., and P. S. Mehringer, Jr. 1965. Pleistocene pollen analysis and biogeography of the Southwest, pp. 433-451. *In* The Quarternary of the United States. Princeton, N. J.: Princeton University Press.

Miller, A. H. 1951. An analysis of the distribution of the birds of California. Univ. Calif. Publ. Zool. 50: 531-644.

Miller, A. H., and R. C. Stebbins. 1964. The lives of desert animals in Joshua Tree National Monument. Berkeley and Los Angeles: University of Calif. Press. 452 p.

Nice, M. M. 1941. The role of territory in bird life. Amer. Midl. Nat. 26:441-487.

Ohmart, R. D. 1969. Dual breeding ranges in Cassin Sparrow (*Aimophila cassinii*), p. 105. *In* C. C. Hoff and M. L. Riedesel, ed., Physiological systems in semiarid environments. Albuquerque: University of New Mexico Press.

Phillips, A., S. Marshall, and G. Monson. 1964. The birds of Arizona. Tucson: University of Ariz. Press. 212 p.

Pohl, H. 1969. Some factors influencing the metabolic response to cold in birds. Proc. Fed. Am. Socs. Exp. Biol. 28:1059-1064.

Pohl, H., and G. C. West. 1973. Daily and seasonal variation in metabolic response to cold during rest and forced exercise in the Common Redpoll. Comp. Biochem. Physiol. 45:851-867.

Rand, A. L., and R. M. Rand. 1943. Breeding notes on the Phainopepla. Auk 60:333-341.

Ricklefs, R. E. 1974. Energetics of reproduction in birds, pp. 152-192. *In* R. A. Paynter, ed., Avian energetics. Nuttall Ornith. Club, Harvard Univ., Cambridge, Mass.

Scott, M. L. 1973. Nutrition in reproduction—direct effects and predictive functions, pp. 46-77. *In* D. S. Farner, ed., Breeding biology of birds. Nat. Acad. Sci., Washington, D.C.

Scott, W. E. D. 1885. On the breeding habits of some Arizona birds, third paper. *Phainopepla nitens.* Auk 2:242-246.

Stiles, F. G. 1973. Food supply and the annual cycle of the Anna Hummingbird. Univ. Calif. Publ. Zool. 97. 109 p.

Thompson, R. D., and C. V. Grant. 1968. Nutritive value of two laboratory diets for Starlings. Lab. Animal Care 18:73-79.

Tucker, V. A. 1968. Respiratory exchange and evaporative water loss in the flying Budgerigar. J. Exp. Biol. 48:67-87.

———. 1973. Bird metabolism during flight: evaluation of a theory. J. Exp. Biol. 58:689-709.

U.S. Environmental Data Service. 1972. Climatological data, California. Vol. 76, no. 13. Asheville, N.C.

———. 1973. Climatological data, California. Vol. 77, no. 13. Asheville, N.C.

U.S. Weather Bureau. 1964. Climatic summary of the United States, supplement for 1951 through 1960. No. 4. Washington, D.C.

Utter, J. M. 1971. Daily energy expenditure of free-living Purple Martins (*Progne subis*) and Mockingbirds (*Mimus polyglottus*) with a comparison of two northern populations of Mockingbirds. Ph.D. dissertation, Rutgers Univ. 173 p.

Utter, J. M., and E.A. LeFebvre. 1973. Daily energy expenditure of Purple Martins (*Progne subis*) during the breeding season: estimates using D_2O^{18} and time budget methods. Ecology 54:597-604.

Wagner, H. O. 1941. Lange "Verlobungzeit" Mexikanischer Tyranniden. Ornithol. Monatsb. 49:137-138.

Walsberg, G. E., 1975. Digestive adaptations of *Phainopepla nitens* associated with the eating of mistletoe berries. Condor 77:169-174.